Growing Microgreens
From Seed To Sale
M.K. Hanna

May Killebrew Hanna

Copyright © 2024 by M.K. Hanna

All rights reserved.

No portion of this book may be reproduced in any form without written permission from the publisher or author, except as permitted by U.S. copyright law.

This publication is designed to provide accurate and authoritative information in regard to the subject matter covered. It is sold with the understanding that neither the author nor the publisher is engaged in rendering legal, investment, accounting or other professional services. While the publisher and author have used their best efforts in preparing this book, they make no representations or warranties with respect to the accuracy or completeness of the contents of this book and specifically disclaim any implied warranties of merchantability or fitness for a particular purpose. No warranty may be created or extended by sales representatives or written sales materials. The advice and strategies contained herein may not be suitable for your situation. You should consult with a professional when appropriate. Neither the publisher nor the author shall be liable for any loss of profit or any other commercial damages, including but not limited to special, incidental, consequential, personal, or other damages.

Book Cover by May Killebrew Hanna

Illustrations and Graphics by May Killebrew Hanna

Contents

Introduction — 1
Why Start A Microgreens Business

1. Understanding The Microgreens Market — 6
2. Developing A Business Plan — 15
3. Setting Up Your Growing Operation — 24
4. Scaling Your Growing Operation — 34
5. Managing Inventory And Costs — 43
   Managing Inventory And Costs
6. Finding Your First Customers — 53
7. Pricing Your Microrgreens — 62
8. Packaging and Branding — 70
9. Expanding Your Product Line — 79
10. Marketing and Promoting Your Business — 88
11. Navigating Challenges And Competition — 99
12. Maintaining Business Growth — 108
13. Sustainable And Ethical Business Practices — 117
14. Microgreens: Varieties, Growing and Harvesting — 125
15. Bonus: Business Tools And Resources — 137
16. Conclusion — 146
    A Path To A Thriving Microgreens Business

17. References 150

Also by 154
M.K. Hanna

Introduction
Why Start A Microgreens Business

The microgreens market is one of the most dynamic and promising sectors in today's food industry. These tiny, nutrient-dense greens have rapidly captured the attention of chefs, health-focused consumers, and sustainability advocates alike. Prized for their vibrant colors, intense flavors, and exceptional nutritional content, microgreens have become staples everywhere, from high-end restaurant plates to everyday home kitchens. As consumer interest surges in local, organic, and health-conscious foods, demand for fresh, responsibly grown microgreens continues to soar—creating an enticing market opportunity for those prepared to meet it.

Brief Overview of the Microgreens Market

Microgreens hold a unique and expanding niche within the agricultural and specialty produce industries. Once exclusively featured in fine dining for their vibrant colors, intense flavors, and striking plate appeal, these miniature greens have since entered the mainstream and are now found in grocery stores, farmers' markets, and specialty health food shops. This rise in popularity is fueled by a growing consumer demand for nutrient-dense foods and a heightened awareness of sustainable farming practices. Today's consumers are increasingly mindful of where and how their food is produced, seeking products that align with their health goals and eco-conscious values. Microgreens are a natural fit for this movement, appealing to shoppers who prioritize freshness, flavor, and environmental impact in their food choices.

For small-scale growers, the microgreens industry offers remarkable business potential. These greens are ideally suited for local niche markets and direct sales to chefs who value fresh, high-quality ingredients to elevate their dishes. They're also popular among farmers' market customers and health-conscious consumers who seek unique and nutritious produce options. From a business perspective, microgreens are highly versatile—they can be marketed in farmers' markets, online through subscription boxes, or offered as a premium product in retail stores. Their relatively low startup costs, quick growing cycles, and high turnover make microgreens a compelling choice for many aspiring entrepreneurs.

The microgreens business offers a manageable and scalable model for those just starting out that can start small and expand as demand grows. And for established small-scale farmers, adding microgreens to their offerings provides a new, profitable revenue stream with significant growth potential. Whether you want to start a side business, diversify an existing operation, or explore sustainable agriculture, the microgreens industry offers a lucrative and rewarding path forward.

Benefits of a Microgreens Business

Starting a microgreens business is not only accessible but can also be highly rewarding. Here are some of the key benefits that make this business venture both achievable and attractive:

- **Low Startup Costs**: Unlike traditional farming, microgreens require only minimal initial investment. Growing indoors or in small greenhouse spaces eliminates the need for large plots of land, and the equipment needed—trays, seeds, and grow lights—is relatively affordable. This low barrier to entry makes it easy for anyone to start growing microgreens on a small scale.

- **Quick Turnover**: Microgreens grow rapidly, with most varieties reaching harvestable size in just one to three weeks. This quick turnaround allows growers to generate revenue quickly, which is ideal for new business owners who need steady cash flow. Because of the short growth cycle, you can experiment with different varieties and respond quickly to customer preferences.

- **Scalability**: Microgreens businesses can start small and scale up as demand grows. What begins as a side project in a spare room or garage can expand into a larger, dedicated space with multiple trays or growing systems. As your customer base grows, you can gradually increase production, add new varieties, or expand into value-added products like microgreen salads and grow kits.

- **Sustainable and Eco-Friendly**: Microgreens are an environmentally friendly crop, requiring less water, land, and energy than traditional crops. This appeals to eco-conscious consumers and growers. By adopting sustainable practices and even exploring organic growing methods, you can align your business with the values of today's health—and environment-focused markets.

Who Is This Book For?

Growing Microgreens from Seed to Sale is designed for a diverse audience—those passionate about fresh, healthy food and looking to turn that passion into a thriving business. Here's a closer look at who will benefit from this book:

- **Aspiring Entrepreneurs**: If you're considering launching your first business, microgreens offer a manageable, affordable, and potentially profitable entry point. This book provides a step-by-step guide to building a business from the ground up, from market research and business planning to pricing, marketing, and customer relationship management.

- **Small-Scale Farmers**: For farmers already operating on a small scale, microgreens can be an excellent addition to an existing product lineup. This book provides insights on how to grow microgreens efficiently and market them alongside other products, helping you diversify revenue streams while meeting the demands of health-conscious customers.

- **Sustainable Living Advocates**: Microgreens fit seamlessly into the ethos of sustainable living, as they require fewer resources and can be grown in small spaces year-round. If you're passionate about eco-friendly practices and want to promote sustainability in your community, this book will show you how to grow and sell microgreens as a sustainable business model.

- **Individuals Seeking a Profitable Side Hustle**: Microgreens offers a unique opportunity if you're looking for a side income that doesn't require a large initial investment. Their short growth cycle and scalability make it possible to manage a small operation alongside a day job, with the flexibility to expand if demand increases.

This book is a practical, comprehensive guide to turning your interest in microgreens into a profitable business. It covers everything you need to know, from selecting the right equipment and understanding market trends to building customer relationships and scaling your operation as demand grows. Whether you're a beginner or have already dabbled in microgreens, this book offers insights, tools, and strategies to take your venture to the next level.

In *Growing Microgreens from Seed to Sale*, you'll find a roadmap for creating a sustainable, successful business that generates income and contributes positively to your community and the environment. With the tools, strategies, and insights in this book, you'll be ready to take the first steps in building a thriving microgreens business uniquely yours.

Chapter 1
Understanding The Microgreens Market

The microgreens industry is rapidly emerging as one of the most dynamic sectors in modern agriculture, fueled by a growing consumer appetite for fresh, nutrient-dense, and sustainably produced foods. Once a specialty item confined to fine dining establishments, microgreens have quickly transitioned into the mainstream, appearing in grocery stores, farmers markets, health food shops, and even home kitchens. This evolution from a culinary trend to a thriving industry presents an exciting opportunity for small-scale growers and aspiring entrepreneurs to tap into a profitable and ever-expanding market.

Today, the demand for microgreens reflects a broader shift in consumer priorities. Health-conscious shoppers are increasingly seeking foods that are not only rich in nutrients but also grown with sustainable practices that align with their environmental values. The rise of farm-to-table dining, the popularity of local and organic foods, and the awareness of food traceability have all contributed to the rapid growth of the microgreens market. Microgreens fit perfectly into this cultural moment: they're compact, resource-efficient, and packed with flavor and nutrition, making them appealing to chefs, home cooks, and health-conscious individuals.

The potential is significant for those looking to enter the industry. Small-scale growers can easily find local markets in restaurants, farmers' markets, and specialty food stores, while others may leverage online platforms to sell directly to consumers. Microgreens are highly versatile in their applications, ranging from garnishes and salad components to smoothies and health-boosting meal additions.

Their short growth cycles, low overhead costs, and scalability make them an ideal crop for entrepreneurs seeking a low-risk yet rewarding business opportunity.

In this chapter, we'll dive into the trends and forces driving the growth of the microgreens industry. We'll explore how to identify and target specific customer segments—from chefs and market-goers to subscription services and health-focused consumers. Finally, we'll examine the niche opportunities within the microgreens space, such as organic, non-GMO, and restaurant-specific varieties, that can help set your business apart in a competitive marketplace. By understanding these foundational elements, you'll be well-prepared to capitalize on the vast potential of the microgreens industry and build a business that thrives in today's food-conscious world.

Overview of Market Trends and Growth Potential

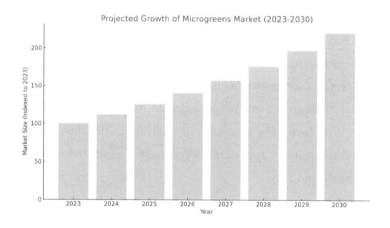

Grand View Research https://www.grandviewresearch.com/industry-analysis/microgreens-market-report

The market for microgreens has experienced rapid growth over the last decade, driven by several key consumer and industry trends. Health-conscious consumers are increasingly looking for nutrient-dense foods, and microgreens have quickly become a favorite for their high concentration of vitamins, minerals, and antioxidants. Studies show that many microgreens contain significantly higher levels of nutrients than their mature vegetable counterparts, making them an appealing choice for customers seeking to add a nutritional boost to their diets.

The rise of sustainable and locally sourced food options has propelled microgreens into the spotlight. Microgreens can be grown in small spaces with minimal resources, requiring less water and energy than traditional crops. This aligns well with the current consumer demand for sustainable, eco-friendly foods. The farm-to-table movement, which emphasizes the use of fresh, local ingredients, has also spurred the growth of microgreens. Many restaurants, especially those emphasizing organic or locally sourced menus, see microgreens as a way to enhance flavor and presentation while supporting sustainability.

The COVID-19 pandemic further accelerated this market as more people became interested in health and wellness and started growing their food. While many began by growing herbs or small vegetables at home, microgreens emerged as an attractive option due to their quick growth cycle and minimal space requirements. Even as restaurants and markets have reopened, the interest in microgreens remains strong, and many small-scale farmers and home-based entrepreneurs are capitalizing on this continued demand.

The global microgreens market is projected to grow significantly over the next decade. With increasing awareness of their health benefits, environmental sustainability, and culinary appeal, microgreens offer a lucrative opportunity for those entering the industry.

Identifying Target Customers

Understanding and defining your target customer base is essential to enter the microgreens market successfully. Identifying the customers most likely to appreciate and buy your product will enable you to tailor your offerings, marketing, and distribution strategies accordingly. Here are some key target customer groups for microgreens businesses:

Chefs and Restaurants

Chefs, particularly those in upscale or health-conscious restaurants, were some of the earliest adopters of microgreens. They value microgreens for their concentrated flavors, vibrant colors, and unique textures, which enhance dishes' taste and presentation. Microgreens such as pea shoots, radish greens, and basil can add distinct flavors and decorative touches that elevate the dining experience.

When targeting chefs, it's important to understand their specific needs and preferences. Many chefs look for consistent, high-quality ingredients and may prefer suppliers who offer unique varieties or flavors that align with their menus. Building relationships with local chefs can be highly beneficial, as repeat business is typical in the restaurant industry. If a chef loves your product and sees it as a reliable and consistent source, they will likely return for regular orders.

Consider crafting a tailored pitch highlighting your microgreens' freshness, flavor, and visual appeal to reach this customer group. Offering sample packs of your varieties can effectively showcase your product's quality and diversity, allowing chefs to experiment with your microgreens before committing to larger orders.

Farmers Markets

Farmers markets are a popular venue for selling microgreens directly to consumers. Customers at farmers' markets are often looking for fresh, local, and organic produce, making it an ideal setting to introduce microgreens. Many people who frequent farmers' markets are also interested in health and nutrition and appreciate microgreens' nutritional value.

Farmers markets also allow you to interact directly with customers, which can be invaluable for building brand loyalty and gathering feedback. For instance, you might find that certain varieties, such as sunflower shoots or arugula microgreens, are especially popular among this crowd. You can also educate customers on using microgreens, sharing recipes, or suggesting pairings, which adds value to the customer experience.

Creating an attractive, well-branded stand is key to drawing attention at farmers' markets. Consider using visually appealing packaging, clear labeling, and informative signage emphasizing your microgreens' health benefits and unique flavors. Offering samples is another excellent way to engage customers and allow them to experience the taste and quality of your product firsthand.

Subscription Services

Subscription boxes and meal delivery services have become popular as more people seek convenient, healthy food options. Microgreens make an ideal addition to these services because of their compact size, nutritional benefits, and short shelf life, which encourage frequent replenishment. You can reach health-conscious customers who appreciate fresh, high-quality ingredients by partnering with a local meal kit or subscription box provider.

For microgreens businesses, offering a subscription service directly can also be a viable option. A microgreens subscription box allows you to provide customers with a rotating selection of microgreens, catering to those who enjoy variety in their diets. For example, customers could receive a weekly or bi-weekly box containing seasonal microgreens or specific varieties that pair well with certain dishes. Subscription services help generate consistent revenue and build a loyal customer base.

Health-Conscious Consumers

Health-conscious consumers are among the most enthusiastic supporters of microgreens, drawn to their high nutrient content and potential health benefits. Many health-focused individuals know that microgreens can provide vitamins, antioxidants, and minerals in a concentrated form, making them a valuable addition to a balanced diet. Microgreens are more than a garnish for these customers—they're a vital part of their nutritional regimen.

Targeting health-conscious consumers may involve selling at health food stores, gyms, wellness centers, or online through e-commerce platforms. Marketing efforts should focus on educating these customers about the specific health benefits of microgreens and their nutritional value. For example, pea shoots are rich in vitamins A and C, while radish microgreens contain high vitamin E and potassium levels.

To attract and retain this customer base, consider creating content highlighting microgreens' health benefits, such as blog posts, social media posts, or informative brochures. Sharing recipes and health tips can also increase engagement and demonstrate the versatility of microgreens in everyday meals.

Niche Opportunities

The microgreens industry offers several niche opportunities that can set your business apart in a competitive market. Focusing on a specific niche allows you to meet specialized demands, differentiate your products, and charge premium prices. Here are some popular niche opportunities within the microgreens market:

Organic Microgreens

As consumer interest in organic produce grows, organic microgreens present a valuable niche opportunity. Organic products are often priced higher, reflecting the additional care and certification processes in organic growing. Pursuing organic certification can be worthwhile for microgreens growers, as it assures customers who prioritize pesticide-free, chemical-free food.

While obtaining organic certification may require additional paperwork, inspections, and fees, the investment can pay off by attracting health-conscious consumers who prefer organic produce. Adopting and promoting organic growing practices can still appeal to customers who prioritize eco-friendly, sustainable foods if full certification isn't an option.

Non-GMO Microgreens

Non-GMO products have gained traction in the health food market, driven by consumers who want to avoid genetically modified organisms in their diets. Offering non-GMO microgreens can be an effective way to reach this audience. By sourcing non-GMO seeds and prominently marketing your product as non-GMO, you can appeal to customers who value natural, unaltered food sources.

Promoting your microgreens as non-GMO can enhance your brand image and help establish trust with your customers, especially in a market where food transparency is increasingly valued.

Restaurant-Specific Varieties

Another lucrative niche opportunity lies in growing microgreens specifically tailored for restaurant menus. Some restaurants, particularly fine dining and farm-to-table establishments, constantly search for unique ingredients that set them apart from competitors. Growing rare or specialty varieties, such as purple basil, shiso, or red cabbage microgreens, can make your product highly desirable to chefs looking for distinctive flavors and colors.

Offering custom-grown microgreens can create a unique selling point that attracts restaurant clients. Discussing chefs' needs and preferences can also lead to exclusive partnerships and bulk sales agreements. Building a reputation for delivering unique, high-quality microgreens can establish your business as a go-to supplier for innovative restaurant chefs.

Microgreen Blends

Blends of microgreens, such as a "salad mix" or "spicy mix," offer a unique selling proposition for customers looking for variety. Combining complementary flavors, textures, and colors allows you to create ready-to-use blends that simplify meal prep for consumers. These pre-made mixes are trendy in retail settings, providing convenience and variety to shoppers.

Blends can be tailored for different uses, such as salads, smoothies, or garnishes. For example, a blend of mild-tasting microgreens (like sunflower and pea shoots) is ideal for salads, while a mix of more peppery greens (like arugula and mustard) works well as a garnish. Offering microgreen blends allows you to differentiate your product and provide added value to customers looking for convenient, versatile options.

Conclusion

Understanding the microgreens market is essential to positioning your business effectively and thriving in a competitive industry. By analyzing the trends fueling the demand for microgreens, pinpointing your ideal customer segments, and exploring specific niche opportunities, you can carve out a unique and sustainable position for your brand. Whether you focus on chefs seeking fresh, unique ingredients, health-conscious consumers who value nutrient-rich foods, or organic shoppers prioritizing sustainable farming practices, a deep understanding of your audience's needs and preferences is critical to long-term success.

Throughout this book, you'll gain the tools, strategies, and insights needed to build a microgreens business that doesn't just meet market demand but also differentiates itself through exceptional quality, innovative offerings, and a commitment to eco-friendly practices. With each chapter, you'll be equipped to make informed decisions about everything from selecting varieties and setting competitive prices to packaging, branding, and marketing your microgreens. By the end, you'll have a comprehensive roadmap to establish a business that resonates with your customers and contributes positively to your community and the environment. A thoughtful approach to market positioning and customer engagement will lead you to growth, profitability, and meaningful impact in an industry ripe with opportunity.

Chapter 2
Developing A Business Plan

A comprehensive business plan is foundational in building a thriving and sustainable business. A well-crafted plan defines your vision, establishes clear, actionable goals, and provides a strategic roadmap for achieving them over time. A business plan is vital for informed decision-making for aspiring entrepreneurs, helping you stay organized, focused, and adaptable as your business evolves.

A business plan serves as more than just a document; it's a framework that clarifies each stage of your business journey, from initial setup to expansion. Setting realistic income targets, mapping out growth objectives, and creating a sound financial strategy gives you the structure needed to manage costs, allocate resources, and meet profitability goals. Additionally, defining your brand identity—how you'll communicate your values and engage with customers—lays the groundwork for building a memorable brand that resonates with your audience.

In this chapter, we'll walk you through the core elements of a business plan, including establishing income and expansion goals, planning finances, selecting the right business structure, and crafting a powerful brand identity. With these essential components in place, you'll be well-prepared to navigate the challenges and opportunities of the microgreens market, setting your business on a path to success and sustainable growth.

Setting Business Goals: Income Targets and Expansion Goals

Setting clear, realistic goals is foundational to any business, and microgreens are no exception. Your goals will serve as benchmarks for progress, guide your decisions, and provide motivation. The two most essential goals to establish are income targets and expansion goals.

Income Targets

Setting income targets is critical to understanding your revenue potential and how much you need to sell to meet financial obligations and personal income expectations. For microgreens, this involves estimating the revenue generated by selling various products, such as individual microgreen trays, packaged microgreens, or value-added items like salads or grow kits.

- **Identify Your Monthly Revenue Goal**: Determine how much you must earn monthly to cover personal and business expenses. For example, if you estimate that you will need $3,000 monthly to break even, this becomes your minimum target.

- **Calculate Product Revenue Requirements**: Consider how much you need to sell to hit this target. If your average microgreen tray sells for $15, you'll have to sell 200 trays monthly to reach a $3,000 income target. Breaking down revenue needs by product allows you to set more precise sales targets.

- **Project Seasonal Variations**: Income targets should account for potential fluctuations in demand due to seasonality. For example, if farmers' markets are only available during warmer months, factor in higher sales during these times and adjust targets for the slower seasons.

Setting income targets ensures you clearly understand the revenue needed to support your business, making it easier to adjust prices, marketing efforts, and product offerings to meet your financial goals.

Expansion Goals

While income targets focus on short-term profitability, expansion goals are about building a foundation for growth. These goals outline how you plan to scale your business, whether by increasing production capacity, adding new product lines, or entering new markets.

- **Establish Production Goals**: Decide how much you want to increase production over time. For instance, your first expansion goal could be to double your weekly output within a year.

- **New Product Development**: Add new varieties or product types to your growth strategy. For example, you could add three new varieties within six months to cater to chefs or develop a ready-to-eat microgreen salad for health food stores.

- **Market Expansion**: Expanding into new sales channels, like partnering with local grocery stores or launching an online subscription box, can broaden your customer base. Determine the timeline and milestones for entering each new market, like partnering with two local restaurants within three months.

By setting income and expansion goals, you'll have a balanced plan that prioritizes immediate financial sustainability while paving the way for future growth.

Financial Planning and Start-Up Costs

Financial planning is the backbone of your business plan, as it defines how you'll fund your start-up costs, manage expenses, and achieve profitability. Microgreens offer a relatively low-cost entry into agriculture, but understanding your expenses and developing a realistic budget is crucial for long-term success.

Initial Start-Up Costs

Estimating start-up costs helps you assess how much capital you'll need to launch your business. For a microgreens operation, start-up costs typically include:

- **Growing Equipment**: Essentials like trays, soil, seeds, and grow lights. A small-scale setup with a few trays and basic equipment might cost $500–$1,000, while a larger setup could range from $1,500–$3,000.

- **Packaging Supplies**: Clear containers, labels, and other materials needed to package your product. Expect to spend around $100–$200 to start.

- **Branding and Marketing Materials**: Costs for business cards, signage for farmers markets, and logo design. Budget around $200–$500, depending on whether you create designs or hire a professional.

- **Legal and Administrative Fees**: These fees can range from $100 to $500 if you register your business, apply for licenses, or create a website.

By listing these initial costs, you can clearly see the funds required to launch your business. If funding is a concern, consider small business grants, low-interest loans, or even a crowdfunding campaign to get started.

Operating Costs

Operating costs cover the recurring expenses involved in running your business. Tracking these costs is essential for setting prices, managing cash flow, and ensuring profitability. Typical microgreens business operating costs include:

- **Seeds and Soil**: These are ongoing costs. Depending on your scale, budget $50–$200 monthly.

- **Water and Electricity**: Energy costs are a consideration if using grow lights or indoor setups. Monthly utility costs may range from $30 to $100, depending on the scale and local rates.

- **Labor Costs**: Labor is an ongoing expense, whether you hire employees or pay yourself. Be realistic about the time required for planting, harvesting, and packaging.

- **Packaging and Distribution**: Costs associated with packaging materials and any distribution methods, such as gas for delivery, can vary depending on your sales volume and customer base.

A clear understanding of your start-up and operating costs will allow you to set accurate pricing and revenue targets, ensuring your business remains financially viable.

Choosing a Business Structure

The structure you choose for your business has significant legal, tax, and financial implications. The two most common business structures for small-scale microgreens operations are sole proprietorship and limited liability company (LLC).

Sole Proprietorship

A sole proprietorship is the simplest business structure and is often the default option for small-scale entrepreneurs. Here are some key characteristics:

- **Simplicity**: A sole proprietorship is easy to establish and requires minimal paperwork. You don't need to register with the state, though some states may require a business license.

- **Tax Benefits**: Taxes are straightforward; all business income is reported on your tax return. However, this structure needs more tax flexibility and potential deductions of an LLC.

- **Unlimited Liability**: With a sole proprietorship, there is no legal separation between personal and business assets, which means you're personally liable for business debts.

A sole proprietorship is often ideal for individuals starting small and looking to keep things simple. However, it's important to consider the potential risks, particularly as your business grows.

Limited Liability Company (LLC)

An LLC is a more formal business structure that offers liability protection and tax benefits. Here's what an LLC entails:

- **Liability Protection**: One of the primary benefits of an LLC is personal asset protection. This means your assets are generally protected from business debts and liabilities.

- **Tax Flexibility**: An LLC allows for pass-through taxation, meaning profits and losses are passed through to your tax return. Alternatively, you can have the LLC taxed as an S Corporation if it's financially advantageous.

- **Professional Image**: Many businesses benefit from an LLC's credibility, especially when working with other businesses, such as restaurants or grocery stores.

While an LLC requires more paperwork and fees than a sole proprietorship, the liability protection and potential tax benefits make it a popular choice for entrepreneurs with growth ambitions.

By choosing the right structure for your business, you can balance simplicity, liability protection, and tax advantages in a way that supports your goals and risk tolerance.

Business Name and Branding Strategies

Your business name and branding are vital to establishing your identity and connecting with customers. A memorable, well-thought-out brand distinguishes your microgreens business from competitors and helps build customer trust.

Choosing a Business Name

A strong business name is distinctive, memorable, and relevant to your product and values. Here are some tips for choosing a name that resonates with your audience:

- **Reflect Your Values**: If sustainability or organic growing is a central part of your business, consider a name that reflects these values (e.g., "GreenLeaf Microgreens" or "Pure Roots").

- **Keep It Simple**: Names that are easy to spell and pronounce are more likely to stick in customers' minds. Avoid complex or lengthy names that could be difficult to remember.

- **Consider Longevity**: Choose a name that allows for future growth. For example, a name like "Urban Greens" may give you flexibility to expand beyond microgreens, whereas "Micro Shoots" is more specific.

Once you've selected a name, check for availability. If you plan to create a website, you'll want to ensure that another business doesn't already register it and is available as a domain name.

Developing a Brand Identity

Your brand identity encompasses everything from your logo and color scheme to the tone of your marketing materials. A cohesive brand helps communicate your business's values, build recognition, and foster customer loyalty. Here's how to start developing your brand:

- **Define Your Brand's Personality**: Determine how you want customers to feel when they think of your brand. Are you going for an eco-friendly, down-to-earth vibe or something more modern and sleek?

- **Choose Colors and Fonts Consistent with Your Image**: The colors and fonts you choose should align with the image you want to convey. For example, green tones and earthy designs often appeal to eco-conscious consumers.

- **Create a Logo**: A logo is the visual centerpiece of your brand. Whether you design it yourself or hire a professional, ensure it's clean, recognizable, and adaptable to various formats (e.g., packaging, signage, social media).

Building Brand Awareness

Once you have a name and brand identity, build awareness through consistent messaging and strategic outreach. Consider social media, business cards, and a website as starting points. Additionally, attending farmers' markets or participating in local events can help you engage with potential customers and establish a memorable presence in your community.

Conclusion

Creating a comprehensive business plan is vital in establishing a successful and resilient microgreens business. By thoughtfully setting income targets and expansion goals, crafting a solid financial strategy, selecting the right business structure, and building a unique brand identity, you lay the groundwork for sustained growth and profitability. This plan is more than a checklist—it's a strategic guide that keeps you focused, adaptable, and prepared to make informed decisions as your business evolves.

With a clear vision and a structured roadmap, you'll be equipped to launch and scale your business confidently, standing out in a competitive market with a strong brand and an efficient operational model. As you implement your plan and witness your goals take shape, your business will grow from an idea into a thriving enterprise that resonates with customers, supports your financial objectives, and contributes positively to the community and environment. By committing to a well-thought-out business strategy, you set your microgreens business on a path to long-term success, positioning yourself as a trusted provider in a high-demand, ever-expanding industry.

Chapter 3
Setting Up Your Growing Operation

B uilding an efficient growing operation is the cornerstone of a successful microgreens business. While microgreens cultivation may appear simple compared to large-scale crop production, setting up a commercial operation that maximizes productivity, maintains consistent quality, and efficiently uses resources requires thoughtful planning and attention to detail. This chapter will guide you through the critical elements of selecting the optimal space, assembling essential equipment, sourcing and managing supplies, and laying the groundwork for a smooth and productive operation.

My previous book, ***Growing Microgreens from Seed to Table***, provides an in-depth look at the growing process for microgreens. In that book, I cover every aspect of cultivation, from selecting the right seeds to mastering harvesting techniques and post-harvest handling, ensuring you achieve high-quality yields. Here, our focus will shift to creating a commercial setup designed specifically for business-scale production, enabling you to meet increasing demand efficiently and sustainably as your business grows. With the proper setup, you'll be well-equipped to deliver fresh, high-quality microgreens to your customers, setting your business apart in a competitive market.

Choosing the Right Space for a Commercial Setup

One of the first decisions you'll need to make is choosing the optimal space for your microgreens business. Unlike larger crops requiring significant land, microgreens can be grown in compact, controlled environments, allowing you to consider various options, including indoor setups, outdoor spaces, greenhouses, or even a home setup. Your choice of space will depend on factors such as budget, climate, production goals, and available infrastructure. Here's a breakdown of each option's benefits and considerations.

Indoor Setup

An indoor setup provides consistent control over growing conditions, making it a popular choice for microgreens growers who prioritize quality and consistency. Indoors, you can control variables like temperature, humidity, and lighting, which are essential to producing high-quality microgreens. Common indoor spaces include basements, spare rooms, or dedicated grow rooms within larger facilities.

- **Advantages**: Indoor setups allow year-round production, regardless of external weather conditions. They're typically safe from pests and environmental contaminants, which can reduce crop loss and improve consistency. Additionally, indoor setups make it easy to stack trays vertically, maximizing production in small spaces.

- **Considerations**: Indoor growing requires an investment in equipment to mimic natural growing conditions, such as grow lights and ventilation systems. Additionally, controlling temperature and humidity can increase electricity costs, which should be factored into your budget.

Indoor Grow Racks With Artificial Lights

Outdoor Setup

Another option is an outdoor setup, such as a backyard or other open area, particularly for growers who prefer natural sunlight and work in mild climates. Outdoor setups allow microgreens to grow under natural conditions, reducing the need for artificial lighting.

- **Advantages**: Outdoor setups can reduce initial costs, as natural sunlight eliminates the need for artificial grow lights. Depending on the area available, they also often offer more space for expansion.

- **Considerations**: Growing outdoors introduces variables like weather, pests, and environmental contaminants. Outdoor setups are less predictable, with changing sunlight and temperature affecting growth cycles and consistency. Seasonality will likely impact production, with slower growth or potential downtime in colder months.

Greenhouse Setup

Greenhouses offer a middle ground between indoor and outdoor setups, combining natural sunlight with the protection and control of an enclosed space. Greenhouses are particularly effective for larger-scale operations that require a semi-controlled environment to increase yields while keeping operating costs manageable.

- **Advantages**: Greenhouses allow for year-round growing, protecting from extreme weather while allowing natural sunlight to reach the plants. They offer more flexibility in managing temperature and humidity and can accommodate larger volumes of microgreens.

- **Considerations**: Greenhouse setups require initial investment in construction and maintenance. Temperature control equipment, such as fans or heaters, may be necessary to manage climate conditions during extreme seasons. Pests can still be a concern in greenhouses, so additional pest management strategies may be required.

Home Setup

Many small-scale growers set up microgreens operations in a home environment, utilizing available spaces like kitchens, garages, or unused rooms. This option works well for those testing the waters or focusing on a limited production capacity.

- **Advantages**: Home setups have low startup costs and minimal operational expenses. They're accessible and easy to monitor, allowing you to experiment with different growing techniques without high overhead.

- **Considerations**: Space constraints can limit production capacity in a home setup, which may require more work to maintain an organized and scalable operation. As the business grows, transitioning to a larger or more dedicated commercial space may be necessary.

Essential Equipment for Microgreens Production

Once you've selected the ideal space, it's time to gather the necessary equipment to ensure efficient and high-quality production. The following are strategic equipment for a commercial microgreens setup to streamline the growing process and improve consistency.

Trays and Containers

Microgreens are typically grown in shallow trays or containers that allow for easy planting, watering, and harvesting. The standard size for trays is 10x20 inches, often called "1020 trays," which fits neatly into most shelving systems. These trays provide a consistent and compact way to manage microgreen crops.

- **Shallow Trays**: Shallow trays provide just enough space for microgreen roots to grow while minimizing soil use and making harvesting easier. Consider using trays with drainage holes to prevent waterlogging.

- **Stackable Shelves**: Stackable shelving systems enable vertical growing, maximizing yield per square foot. They're ideal for indoor setups, allowing you to layer trays under grow lights and irrigation systems.

10 x 20" Growing Trays With Solid Bottom

Grow Lights

Proper lighting is crucial to growing healthy, vibrant microgreens, particularly indoors with limited or inconsistent natural light. LED grow lights are popular due to their energy efficiency and customizable light spectrum.

- **LED Lights**: LED lights offer efficient, low-heat lighting that is adjustable to specific wavelengths needed for plant growth. They're also cost-effective in terms of energy usage, which helps keep operating costs down.

- **Fluorescent Lights**: Fluorescent lights are another option with lower upfront costs than LEDs. However, they're less efficient in the long term and may require more frequent replacement.

Irrigation Systems

Consistent moisture is essential for microgreens, as their shallow roots can quickly dry out. Depending on the scale of your operation, irrigation can be managed manually with a spray bottle or watering can or automated for large-scale production.

- **Manual Watering**: For small operations, manual watering with a spray bottle is often sufficient and offers precise control over moisture levels.

- **Automated Misting Systems**: Automated misting systems help maintain consistent moisture levels across trays for larger-scale setups. These systems distribute a fine mist that prevents overwatering and promotes uniform growth.

Seeds

High-quality seeds are foundational to producing flavorful, healthy microgreens. The choice of seeds will depend on your target market and the specific flavors, textures, and colors you want to offer.

- **Microgreens-Specific Seeds**: Purchase seeds specifically labeled for microgreens, as these are often tested and optimized for higher yields and faster germination.

- **Certified Organic or Non-GMO Seeds**: Many consumers seek organic or non-GMO microgreens, so sourcing seeds that meet these standards can add value to your product. Purchasing seeds in bulk from reputable suppliers can reduce costs and ensure a steady supply.

Sourcing and Managing Supplies

Consistent and reliable supply sourcing is essential to maintaining a smooth operation. Here's how to source and manage the most important materials for your microgreens business.

Seeds

Choosing a reliable seed supplier is essential to ensuring consistent quality and availability. Look for suppliers specializing in microgreens or sprouts, as these vendors often provide seeds tested for germination rates and quality.

- **Consider Bulk Purchases**: Buying in bulk can lower costs per unit and reduce the need for frequent reordering. Many suppliers offer bulk pricing, which is beneficial as your business grows.

- **Regular Supplier Assessment**: Cultivate relationships with multiple suppliers to mitigate supply chain disruptions and ensure competitive pricing.

Soil and Growing Mediums

While soil is the traditional medium for growing microgreens, other mediums, such as coconut coir, hemp mats, and compostable pads, can also work well. Your choice will depend on your setup and any specific needs of your microgreens.

- **Organic or Nutrient-Enriched Soil**: Organic soil or soil enriched with essential nutrients is often preferred. It provides a clean growing environment without synthetic additives, which can appeal to health-conscious consumers.

- **Grow Pads or Mats**: For growers who want to avoid soil altogether, mats made from materials like coconut coir or hemp can be a sustainable alternative. They offer a clean, pest-free environment and are easily disposable or compostable.

Packaging Materials

Thoughtful packaging protects your microgreens and enhances your brand. Choose packaging that keeps your product fresh, visually appealing, and aligned with your brand's image.

- **Clamshell Containers**: Clear plastic clamshells are widely used for packaging microgreens, offering protection and visibility for the product. To appeal to environmentally conscious consumers, look for eco-friendly options, such as recycled or compostable materials.

- **Branded Labels**: Labels are essential for branding and compliance, especially if marketing organic or non-GMO products. Your labels should display the product name, your brand, and any certifications.

Inventory Management

Efficient inventory management is crucial for controlling costs and minimizing waste. Implementing an inventory tracking system, whether manual or digital, allows you to monitor stock levels, track costs, and schedule reordering.

- **Manual Tracking Systems**: For small operations, a spreadsheet may be sufficient to track inventory, including seeds, soil, trays, and packaging.

- **Inventory Software**: As you scale, consider investing in inventory software that can automate tracking, alert you when supplies run low, and help manage costs efficiently.

Conclusion

Setting up a successful commercial microgreens operation requires strategic planning and wise investments in the right space, equipment, and supplies.

Every decision—from choosing the ideal growing space to selecting reliable, efficient equipment and managing inventory—significantly shapes your business's productivity, efficiency, and overall success. Taking time to design your setup thoughtfully optimizes your production capacity. It ensures that you deliver consistently fresh, high-quality microgreens to your customers, meeting market demand and establishing your reputation.

Whether working with a small indoor setup, a greenhouse, or a larger outdoor operation, each choice should align with your production goals and long-term vision. A streamlined, well-organized growing operation allows you to work more efficiently, control costs, and create a sustainable growth model that scales smoothly with demand.

For readers new to microgreens or those who want to sharpen their growing expertise, my previous book, ***Growing Microgreens from Seed to Table***, offers a detailed, step-by-step guide to the cultivation process, from germination and seed selection to harvest and post-harvest handling as well as over fifty recipes using microgreens. These resources will equip you with the technical knowledge and strategic insights needed to set up, and grow microgreens precisely and confidently. With the right foundations in place, you'll be well-prepared to navigate the challenges and opportunities of commercial production, building a brand that stands out for quality and consistency.

Chapter 4
Scaling Your Growing Operation

As demand for your microgreens grows, scaling up your operation from a small-scale setup to a full-fledged commercial enterprise can be rewarding and essential. Expanding your production allows you to reach new customer segments, increase revenue, and position your brand as a trusted supplier in the market. But scaling up successfully requires more than simply adding extra trays and lights. A more extensive operation demands strategic adjustments from space optimization and equipment upgrades to labor management and quality control.

Growing from a small setup to a commercial operation presents unique challenges, as each step in the process must be tailored for efficiency, quality, and consistency. Adopting techniques that maximize yield, streamline production, and support sustainable growth is essential to ensure your business can meet higher demand while maintaining the standards your customers expect. Each decision, from choosing the right watering system to optimizing your lighting setup, impacts your business's productivity and reliability.

In this chapter, we'll explore the essential changes to consider when scaling up, focusing on maximizing yield efficiency through optimized space, watering, and lighting systems. You'll also learn how to expand your offerings by growing a variety of microgreens that appeal to diverse customers, from chefs to health-conscious consumers. Finally, we'll cover methods for maintaining high-quality crops that guarantee customer satisfaction and encourage repeat sales. By understanding these foundational elements, you'll be well-prepared to confidently grow your operation and position your business for long-term success in a competitive market.

From Small Scale to Large Scale: What Changes?

Starting small offers an excellent opportunity to learn the ins and outs of microgreens cultivation. A small-scale setup typically requires minimal investment, making it feasible to experiment with growing techniques, identify preferred varieties, and build an initial customer base. However, when you're ready to expand, some aspects of your operation will need to change to support higher volumes and broader distribution.

Space Requirements

Small scale, a spare room, or even a well-lit kitchen counter might be enough for a few trays. But as you scale up, you'll need a dedicated growing space with ample room for expansion. Moving into a larger indoor area, greenhouse, or outdoor setup allows for vertical stacking, optimal airflow, and lighting placement, all essential for higher production.

- **Indoor Expansion**: If expanding indoors, consider investing in shelving systems and vertical grow racks that allow for efficient use of space and layered production.

- **Greenhouse or Outdoor Setup**: For larger-scale operations, a greenhouse provides a semi-controlled environment that combines natural sunlight with temperature and humidity regulation. Outdoor growing, while exposed to environmental variables, can offer significant space for growth and may reduce lighting costs.

Equipment and Infrastructure

Simple tools like spray bottles for watering and a single grow light may suffice in small-scale operations. However, larger-scale production requires more advanced equipment to efficiently handle watering, lighting, and airflow.

- **Automated Irrigation Systems**: For consistent watering across many trays, automated systems like misting or drip irrigation reduce labor and ensure each tray receives the right amount of moisture.

- **Industrial Grow Lights**: Larger setups benefit from high-quality LED grow lights or full-spectrum fluorescent lights that cover greater surface areas and optimize growth conditions.

- **Air Circulation and Climate Control**: As you add more trays, airflow becomes essential to prevent mold and support healthy growth. Fans and, if indoors, dehumidifiers can help maintain optimal air quality.

Spacious Indoor Grow Room

Labor and Time Management

Scaling up introduces new labor requirements, as more trays mean more time spent planting, watering, harvesting, and packaging. Hiring additional staff or implementing a clear workflow system will help maintain productivity and ensure that quality standards are met as production scales.

- **Delegating Tasks**: As production increases, breaking down tasks into planting, harvesting, and packaging roles can improve efficiency.

- **Scheduling**: Use a calendar or digital planner to create a schedule for planting, watering, and harvesting, which is particularly helpful when managing multiple trays and crop rotations.

Maximizing Yield Efficiency

Optimizing the use of space, water, lighting, and other resources is essential to ensuring profitability and reducing waste as you scale. Maximizing yield efficiency enables you to meet higher demand without excessively increasing costs, contributing to a sustainable business model.

Optimizing Space

Efficient use of space is key to maximizing yield. Implementing vertical shelving units or grow racks allows you to stack trays and increase the growing area without requiring additional floor space.

- **Vertical Stacking**: Multi-tiered shelving systems enable you to grow multiple layers of microgreens, with each shelf supported by dedicated lighting and irrigation systems. Vertical stacking works especially well for indoor or greenhouse setups.

- **Tray Spacing**: Maintaining a balanced spacing between trays ensures airflow, reduces the risk of mold, and allows each tray to receive even light distribution.

Watering Techniques

Watering is critical to yield; too much water can lead to root rot, while too little water can stunt growth. As you scale up, managing watering efficiently becomes more complex and essential to maintaining healthy crops.

- **Automated Misting Systems**: Automated misting systems distribute water evenly, reduce manual labor, and maintain consistent moisture levels. These systems can be timed to mist trays at specific intervals, ensuring optimal hydration.

- **Bottom Watering**: Bottom watering—where water is placed in a tray below the soil layer—encourages roots to grow downward and minimizes the risk of mold on the surface. This technique works well with automated or semi-automated irrigation systems.

Efficient Lighting

Proper lighting is critical for healthy growth, and scaling up means optimizing light placement and timing to cover larger areas. LED lights are particularly effective for energy-efficient, scalable lighting.

- **Adjustable LED Lighting**: LED lights provide a full spectrum that can be adjusted to specific plant needs, supporting robust growth while reducing energy consumption. Position lights at a height that distributes light evenly across each tray.

- **Timed Lighting Cycles**: Establishing a set lighting schedule, typically 12–16 hours on and 8–12 hours off, mimics natural daylight cycles and enhances growth while keeping energy usage consistent.

Growing Multiple Varieties for Diverse Clientele

Expanding your operation presents an opportunity to cater to a wider range of customers by offering multiple varieties of microgreens. Diverse offerings can attract different customer segments, including chefs, health-conscious individuals, and specialty stores, each of whom may be looking for unique flavors, textures, and visual appeal.

Selecting Varieties

Choosing which varieties to grow should be guided by market demand, customer feedback, and the practical requirements of each crop. Popular microgreens like arugula, radish, and sunflower are often in high demand, but consider growing lesser-known varieties to appeal to niche markets.

- **Flavor Profiles**: Offering a mix of flavors—from spicy radish and mustard greens to mild sunflower shoots—provides variety for customers seeking to add flavor complexity to dishes.

- **Visual Appeal**: Microgreens with vibrant colors, like purple basil or red cabbage, attract chefs looking to enhance the visual presentation of their dishes. Consider maintaining a balance of green and colorful microgreens to serve health-focused and aesthetic-minded customers.

- **Nutritional Value**: Health-conscious customers often seek specific nutritional benefits, so growing nutrient-dense varieties like broccoli, which is high in vitamins C and K, can increase appeal among wellness-oriented consumers.

Market Research and Testing

Before dedicating significant space to new varieties, consider conducting market research to identify which microgreens have the most demand. Testing small batches allows you to gauge customer interest and fine-tune growing techniques for each variety.

- **Customer Surveys**: Gathering feedback from existing customers at farmers' markets or online can provide insight into which varieties to add or expand upon.

- **Trial Runs**: Growing small batches of new varieties helps you identify specific care requirements and assess potential profitability before committing to large-scale production.

Maintaining High-Quality Crops for Consistent Sales

As you scale up, maintaining the quality of your microgreens becomes both a priority and a challenge. Consistent quality is critical to building a loyal customer base, as customers depend on you for fresh, flavorful, and visually appealing microgreens.

Regular Quality Control Checks

Implement routine checks for factors like growth rate, color, and taste to ensure consistency across all trays. Consistent monitoring and record-keeping help you identify issues early and maintain high standards.

- **Growth Monitoring**: Track growth rates by documenting each batch's planting, germination, and harvesting times. This allows you to adjust for inconsistencies and keep all trays on a standard growth schedule.

- **Uniform Lighting and Watering**: Regularly check your lighting and irrigation systems to ensure even coverage across trays. Uneven lighting or watering can lead to discrepancies in quality and yield.

- **Taste Testing**: Conduct periodic taste tests to ensure the flavor profile remains consistent and meets customer expectations. Subtle changes in taste may signal soil, watering, or seed quality issues.

Hygiene and Sanitation Practices

Maintaining a clean, organized growing environment is essential to preventing contamination and supporting healthy growth. Scaling up production makes hygiene protocols even more important, as contamination can spread more easily when handling larger volumes.

- **Sanitize Trays and Equipment**: Regularly clean and sanitize trays, cutting tools, and irrigation systems to prevent mold, pests, and disease. Use food-safe sanitizers or diluted hydrogen peroxide as appropriate.

- **Implement Pest Control Measures**: Pests can pose a risk to larger-scale operations, so consider implementing physical barriers like netting or sticky traps to keep them away from growing areas. Natural pest control options can be effective for greenhouse or outdoor setups.

Packaging and Post-Harvest Handling

High-quality post-harvest handling preserves freshness and extends the shelf life of your microgreens, an essential aspect of meeting customer expectations and securing repeat business.

- **Cold Storage**: If possible, store harvested microgreens in a cool, humid environment to retain freshness and prevent wilting. A walk-in cooler or refrigerator can prolong shelf life.

- **Packaging**: Invest in quality packaging that prevents damage, retains moisture, and enhances presentation. Consider eco-friendly, compostable packaging options to appeal to environmentally conscious customers.

Conclusion

Scaling your microgreens operation from a small-scale setup to a robust commercial venture requires thoughtful planning, strategic investments, and a commitment to refining every aspect of your growing process. As you expand, space, equipment, and workflow adjustments are essential for maximizing productivity, maintaining quality, and ensuring consistency in every batch. By focusing on yield efficiency, introducing diverse microgreen varieties, and implementing rigorous quality control, you'll be ready to meet increased demand without compromising the quality your customers expect and value.

Whether catering to chefs seeking unique flavors, health-conscious consumers interested in fresh, nutrient-rich greens, or farmers market shoppers eager for locally grown produce, a well-optimized and efficient operation positions your business for success. As you scale up, each improvement supports higher output and strengthens your brand's reputation for reliability, quality, and innovation in a competitive market.

With careful planning and a dedication to excellence, scaling your operation can lead to long-term profitability, unlocking new customer segments and opportunities. By establishing a solid foundation in efficient practices, thoughtful crop diversity, and exceptional quality standards, your microgreens business can thrive as a respected and trusted name in the industry. In this way, scaling isn't just about growth—it's about building a sustainable brand that stands out for its commitment to delivering the very best microgreens to an expanding market.

Chapter 5
Managing Inventory And Costs

Managing Inventory And Costs

Effective inventory and cost management form the backbone of a profitable and sustainable microgreens business. Every detail counts in a high-turnover industry where freshness, quality, and consistency are paramount. Each operational element is critical in controlling costs and maintaining product quality, from optimizing your harvest schedule to managing storage and reducing waste. Without a structured approach to these areas, even a thriving business can face challenges like unnecessary expenses, product loss, or reduced shelf life, which can impact profitability and customer satisfaction.

Inventory management for microgreens involves careful planning to ensure a steady supply of fresh products, avoid overproduction, and reduce waste. On the other hand, cost management requires a deep understanding of fixed and variable expenses, including seeds, labor, overhead, and packaging. By refining your practices in these areas, you can streamline operations, minimize waste, and maintain the quality and freshness that keeps customers returning.

In this chapter, we'll dive into the essential aspects of managing your inventory and costs effectively. You'll learn how to establish a precise harvest schedule and rotation plan, keep detailed records of expenses, extend your product's shelf life with proper storage techniques, and incorporate sustainable practices that reduce waste and maximize resource efficiency. By implementing these strategies, you'll set up your business for stable growth, improved profitability, and a reputation for quality and responsibility in a competitive market.

Harvest Scheduling and Rotation

An organized harvest schedule and crop rotation system are vital to maintaining a steady supply of fresh microgreens for your customers. Without careful planning, it's easy to fall behind on planting, misjudge harvest timing, or struggle with inconsistent yields—all of which can affect sales and customer satisfaction.

Setting Up a Harvest Schedule

Microgreens have short growth cycles, typically taking anywhere from 7 to 21 days to mature, depending on the variety. Establish a rotating harvest schedule that aligns with your demand to ensure you always have fresh product available. This schedule should be designed to meet regular customer orders, such as weekly orders from restaurants or markets, as well as any expected increases in demand during peak seasons.

1. **Determine Growth Times by Variety**: Start by identifying the growth time for each variety of microgreens you produce. Fast-growing varieties like radish or mustard may only take 7–10 days, while others like sunflower or pea shoots may require closer to 14–21 days.

2. **Create a Planting Calendar**: Once you know the growth times, create a calendar specifying each crop's planting and harvest dates. This calendar should account for both weekly planting rotations and anticipated demand spikes. For instance, planting a new batch of sunflower shoots every Monday may ensure a steady weekly harvest.

3. **Stagger Planting for Consistent Yield**: To maintain a continuous supply, stagger plantings so that you have multiple batches at different stages of growth. If demand increases suddenly, you can adjust by adding an extra planting session to your schedule.

Crop Rotation for Resource Efficiency

Rotating crop varieties helps balance the nutrient demand on your growing medium and reduces the risk of pests and diseases. While microgreens typically don't require crop rotation in the same way as field crops, alternating between different types of microgreens can improve soil health and maintain steady yields.

1. **Alternate Varieties**: Instead of planting the same crop repeatedly, try rotating between fast-growing and slower-growing varieties. This can help prevent nutrient depletion and improve yields across different microgreens.

2. **Monitor and Adjust**: Monitor each crop's performance on a rotation schedule and make adjustments as needed. By observing yield trends, you can determine whether a specific rotation pattern is beneficial for your operation.

Keeping Track of Costs

Accurate cost tracking is fundamental to maintaining profitability in a microgreens business. Understanding your expenses—such as seeds, supplies, labor, and overhead—allows you to make informed pricing decisions, optimize your budget, and identify areas for potential cost savings.

Seed and Supply Costs

Seeds and supplies are ongoing expenses and can account for a significant portion of your operating costs. Regularly tracking these expenses helps you control your budget and make bulk purchases when appropriate.

- **Track Seed Costs by Variety**: Each microgreen variety has a unique cost per seed, depending on factors like seed density and yield. Tracking these costs by variety helps you estimate the expense of each batch, ensuring that high-cost varieties are priced accordingly.

- **Purchase in Bulk for Savings**: Purchasing seeds and supplies in bulk can reduce costs. Work with reputable suppliers to buy larger quantities at a discount, but balance this with your storage capacity to avoid spoilage or waste.

- **Log Supply Usage**: Track other supplies, such as soil, trays, and packaging, by logging usage rates. This data allows you to project how frequently you'll need to reorder supplies, aiding in budget planning and inventory management.

Labor Costs

Labor costs include the time spent on planting, watering, harvesting, packaging, and distribution. Whether you work alone or hire employees, tracking labor costs is essential for understanding your profit margins.

- **Calculate Hourly Labor Costs**: Determine the hourly labor cost for each production step, including planting, harvesting, and packaging. For instance, if harvesting and packaging take two hours per batch and your labor cost is $15 per hour, that batch costs $30 in labor alone.

- **Track Employee Hours**: If you have employees, maintain accurate records of their work hours and tasks. This will help you calculate labor costs more precisely and identify areas where workflow efficiency can be improved.

Overhead Expenses

Overhead expenses include rent, utilities, equipment maintenance, and other fixed costs not directly tied to production but are necessary for running the business. These costs should be included in your cost-per-unit calculations to ensure accurate pricing.

- **Allocate Overhead per Unit**: Calculate your monthly overhead costs and divide this figure by the number of units you produce each month. For example, if your overhead is $500 monthly and you produce 1,000 trays, your overhead cost is $0.50.

- **Review Overhead Regularly**: Overhead costs may fluctuate based on utility rates or rent changes. Regularly reviewing these expenses helps you align your pricing with actual operating costs.

Storage and Shelf-Life Management

Effective storage practices are essential for extending the shelf life of your microgreens and ensuring that they reach customers in peak condition. Since microgreens are perishable, even slight improvements in storage techniques can make a big difference in freshness and reduce waste.

Optimal Harvest Timing

Harvesting at the right time preserves freshness and maximizes shelf life. Aim to harvest microgreens when they're fully mature but before they start to wilt or lose color. This is especially important for high-demand varieties or when preparing for a busy market day.

1. **Monitor Growth Stages**: Pay attention to each batch's growth stage. Microgreens should be harvested when they're at their optimal size and flavor.

2. **Harvest in the Morning**: Harvest in the morning when temperatures are cooler, as this helps retain freshness and reduce wilting.

Post-Harvest Handling

Careful handling of microgreens immediately after harvest is crucial for preserving their quality. Proper handling minimizes damage, reduces spoilage, and extends shelf life.

- **Cool Immediately**: Place harvested microgreens in a cooler or refrigerate immediately after harvesting to slow respiration and maintain freshness.

- **Handle with Clean Tools**: To prevent contamination, use sanitized tools and surfaces during harvesting and packaging. Even minor exposure to contaminants can affect the shelf life of your product.

Storage Conditions

Proper storage conditions help maximize shelf life and maintain quality. Microgreens are best stored in cool, humid environments to prevent wilting and retain their vibrant appearance.

- **Refrigeration**: Keep microgreens in a refrigerated environment between 34°F and 38°F, optimal for maintaining freshness.

- **Humidity Control**: Microgreens can quickly lose moisture, so storing them in containers that maintain adequate humidity prevents wilting. Clamshell containers are effective for maintaining moisture while providing a protective layer.

Packaging for Freshness

Packaging is essential in protecting microgreens and extending their shelf life. Choose packaging that provides visibility for customers while preventing damage during transport.

- **Eco-Friendly Packaging**: Using biodegradable or compostable containers can appeal to environmentally conscious customers while keeping your product fresh.

- **Ventilated Containers**: Ensure packaging has adequate ventilation to allow for air circulation, which helps reduce moisture buildup and prolongs freshness.

Waste Reduction and Sustainability Practices

Reducing waste and implementing sustainable practices are environmentally responsible and can improve profitability by maximizing resource use. Here are some effective strategies for managing waste and enhancing sustainability in your microgreens operation.

Minimize Product Waste

Managing product waste involves preventing crop loss, reducing unsold inventory, and finding ways to repurpose or recycle excess microgreens.

- **Harvest Based on Demand**: By aligning your harvest schedule with customer demand, you can avoid producing more than you can sell. Keep track of sales trends and adjust planting rotations accordingly to match demand.

- **Use Unsold Microgreens in Value-Added Products**: If you have leftover microgreens after a market, consider creating value-added products like microgreen salads, smoothie packs, or dried microgreen seasonings to reduce waste and add another revenue stream.

- **Donate Excess Product**: Partnering with local food banks or community kitchens to donate excess microgreens can help reduce waste and support your community.

Reuse and Recycle Growing Materials

Sustainability practices should extend to the materials used in your growing operation. Reusing and recycling growing trays, soil, and other supplies reduces waste and can lower costs over time.

- **Reuse Growing Trays**: Properly sanitize and reuse growing trays to reduce the need for replacements. Regularly inspect trays for signs of wear to ensure they're still effective.

- **Compost Organic Waste**: Compost any unused soil, stems, and other organic waste to create nutrient-rich compost for future use. This reduces waste and contributes to a more circular growing process.

Implement Energy-Efficient Practices

Energy-efficient practices help lower overhead costs and minimize environmental impact. These small adjustments can improve your bottom line and your business's sustainability.

- **LED Grow Lights**: LED lights use significantly less energy than traditional lighting options and last longer, making them ideal for sustainable microgreens operations.

- **Timers and Sensors**: Installing timers for lights and irrigation systems helps reduce energy use by limiting equipment operation to only when needed, contributing to cost savings and sustainability.

Sustainable Packaging Solutions

Packaging is a major contributor to waste, so adopting eco-friendly options can benefit both your business and the environment. Customers appreciate environmentally conscious packaging and are more likely to support a business with sustainable practices.

- **Biodegradable Containers**: Using compostable or biodegradable containers minimizes waste and can enhance your brand image with eco-conscious customers.

- **Recyclable Labels and Branding**: Choose labels that are recyclable or printed with eco-friendly ink, furthering your commitment to sustainable practices.

Conclusion

Effective inventory and cost management are essential pillars for sustaining profitability and promoting long-term growth in your microgreens business. As your operation expands, having a well-organized harvest schedule, a precise approach to tracking expenses, optimized storage methods, and a commitment to reducing waste will enable you to create a streamlined, resilient, and sustainable business. Each of these practices supports financial stability and enhances the quality of your product, ensuring that customers receive fresh, high-quality microgreens every time.

By meticulously planning and refining these operational elements, you can reduce unnecessary costs, minimize product loss, and build a business model that values both efficiency and sustainability. Implementing eco-friendly practices, such as reducing waste through efficient crop management and choosing sustainable packaging options, allows your business to resonate with today's environmentally conscious consumers looking to support brands with responsible practices.

Through consistent attention to detail and a commitment to continuous improvement, these strategies will set your business apart in a competitive market, establishing you as a reliable, quality-focused, and eco-conscious supplier. Effective inventory and cost management are not just about controlling expenses—they are about creating a business model that reflects your values, builds customer trust, and positions your microgreens business for sustainable success and growth in the years to come.

Chapter 6
Finding Your First Customers

Finding your first customers is one of the most rewarding yet challenging steps in launching a microgreens business. Attracting a loyal customer base goes beyond simply growing a high-quality product; it's about strategically positioning your brand and effectively reaching the right audience. The journey to securing those initial customers requires a mix of persistence, creativity, and a deep understanding of what makes your product unique. Each customer connection is a valuable opportunity to introduce your microgreens and share the story behind your business, highlighting the passion, purpose, and dedication that sets you apart.

There are numerous avenues to explore, from approaching chefs and setting up eye-catching displays at farmers' markets to establishing an online presence and forming partnerships with specialty stores. Each method offers a unique platform for showcasing your microgreens, whether it's the quality that chefs seek, the local touch that farmers' market shoppers appreciate, or the convenience that online customers value. Successfully tapping into these diverse channels broadens your customer base and fosters meaningful relationships that will help your business grow sustainably.

This chapter will guide you through essential strategies for finding and securing your first customers, from crafting the perfect pitch for restaurant partnerships to setting up engaging farmers' market stands and building a compelling online store. We'll also cover ways to cultivate strong, lasting customer relationships, creating a brand experience that encourages repeat purchases and turns one-time buyers into loyal supporters. As you move forward, remember that building a customer base is more than a single transaction—it's about creating an experience and fostering connections that bring people back, time and again, to your business.

Approaching Chefs and Restaurants: Crafting the Perfect Pitch

Chefs and restaurant owners can be some of the most valuable first customers for a microgreens business, as they are always looking for fresh, high-quality ingredients to elevate their dishes. However, establishing these connections requires a strategic approach, as chefs often have specific quality, availability, and consistency requirements. Here's how to approach them effectively:

Understanding What Chefs Look For

To craft a compelling pitch, it's essential to understand what chefs prioritize when sourcing ingredients:

- **Quality and Freshness**: Consistently fresh, vibrant microgreens are crucial for chefs, who use them to add flavor and presentation to dishes. Emphasize your product's high quality and freshness and how quickly it goes from harvest to delivery.

- **Unique Varieties**: Specialty varieties like red amaranth, pea shoots, or purple radish microgreens can set you apart. Chefs appreciate unique flavors, colors, and textures, so be prepared to highlight any distinctive offerings.

- **Consistency and Availability**: Chefs rely on dependable suppliers, so emphasize your ability to provide consistent quality and discuss how you manage potential fluctuations in supply, especially if seasonality affects certain varieties.

Crafting Your Pitch

A well-crafted pitch is clear, concise, and tailored to the specific needs of each chef or restaurant. Here's a structure to guide you:

1. **Introduction**: Briefly introduce your business, focusing on what sets you apart. Mention your commitment to quality, sustainability, or unique varieties if they align with the restaurant's values.

2. **Highlight Benefits**: Explain how your microgreens can enhance the chef's dishes, focusing on the fresh flavor, unique aesthetics, and nutritional value. Describe how your microgreens can complement their menu or add value to specific dishes.

3. **Offer a Sample**: Samples are essential; chefs need to see and taste the quality firsthand. Bring a sample pack of various microgreens and suggest pairing options for their dishes, making it easy for them to imagine incorporating your product.

4. **Suggest a Partnership**: Be prepared to discuss pricing and consistency. Consider offering a special discount on the first order to encourage a trial. Discuss the potential for a lasting partnership in which you supply fresh microgreens on a regular basis.

5. **Call to Action**: End with a clear next step, such as arranging a follow-up meeting, tasting session, or a scheduled delivery of a sample order.

Building relationships with chefs may take time, but persistence and a customer-focused approach will lay the groundwork for a fruitful partnership. Personal touches, such as follow-up emails after meetings or invitations to visit your farm, help establish trust and familiarity.

Farmers Markets and Other Local Outlets: Setting Up a Stand and Branding Tips

Farmers markets are a prime opportunity to connect directly with customers, gain valuable feedback, and establish your brand's presence in the community. Many farmers' market consumers are specifically looking for fresh, local, and organic products, making it an ideal setting for introducing your microgreens.

Setting Up an Eye-Catching Stand

A well-organized, visually appealing stand draws in curious customers. Here are a few tips for creating a stand that stands out:

- **Use Clean, Clear Signage**: Display clear signage that includes your business name, logo, and a brief tagline (e.g., "Fresh, Local Microgreens"). This will help passersby understand who you are and what you offer at a glance.

- **Create a Welcoming Display**: Arrange your microgreens attractively, with plenty of space for customers to browse. Use baskets, trays, or stands to create height and visual interest, and consider including labels for each microgreen variety that list their flavor profile or suggested uses.

- **Offer Samples**: Samples are a great way to introduce customers to the taste and texture of your microgreens. Set up a small, hygienic sample station and offer tasting cups or sample plates to encourage customers to try different varieties.

Branding Tips for Farmers Markets

Branding is essential at farmers' markets, as it helps customers remember you and encourages repeat business. Here's how to make your brand memorable:

- **Consistent Design Elements**: Use consistent colors, fonts, and logo designs on all your marketing materials, from business cards to banners. A cohesive look reinforces brand recognition.

- **Share Your Story**: Customers love to connect with the story behind the product. Display a small sign with your story, mission, or the sustainable practices you follow. Customers who understand your values are more likely to feel connected to your brand.

- **Collect Contact Information**: Set up a mailing list sign-up or offer a small discount for customers who join your email list. Building a customer base at the farmers market can help you stay in touch, share news, and announce future market appearances.

Specialty Foods and Health Food Stores

Specialty food stores and health food markets are excellent outlets for microgreens, especially if you can highlight their health benefits and premium quality. These stores often attract customers who value fresh, local produce and are willing to pay for high-quality items.

Approaching Store Managers

When reaching out to store managers, prepare a pitch highlighting how your product fits into their store's mission and appeals to their customers. Here's a guide to approaching them effectively:

1. **Research the Store's Values**: Understand the store's values and customer demographics. If they focus on organic, local, or health-focused products, emphasize how your microgreens align with these themes.

2. **Highlight Unique Selling Points**: Emphasize what makes your microgreens stand out—your sustainable growing practices, vibrant varieties, or freshness.

3. **Provide Samples and Information**: Provide a small sample package and a brochure or fact sheet on your microgreens, including nutritional information and suggested uses.

4. **Discuss Restocking and Delivery**: Specialty stores need consistent restocking, so be prepared to discuss delivery schedules and how you'll ensure a steady supply. Offering fresh deliveries multiple times a week may appeal to stores focused on freshness.

Selling Online: E-Commerce Platforms and Subscription Boxes

Selling online can help you reach a wider audience beyond your local market, offering the convenience of direct-to-door deliveries. Building loyalty and trust in online sales calls for a compelling brand identity, seamless user experience, and ongoing customer engagement.

Choosing an E-Commerce Platform

Selecting the right e-commerce platform can simplify your sales and ensure a positive customer experience. Here are some popular options:

- **Shopify**: Shopify offers a range of features tailored for small businesses, from customizable product pages to secure payment options and shipping integrations. It's user-friendly and ideal for selling fresh products like microgreens.

- **Etsy**: Known for artisan and specialty products, Etsy is an excellent platform for businesses focusing on organic, handmade, or unique offerings. Etsy's audience may be particularly interested in local, sustainably grown produce.

- **WooCommerce**: If you already have a WordPress site, WooCommerce integrates seamlessly to add an online store feature. It's highly customizable, allowing you to tailor your store to your brand's look and feel.

Offering Subscription Boxes

Subscription boxes are a great way to encourage repeat purchases and build a loyal customer base. Here's how to create a successful subscription model:

- **Define Your Box Contents**: To add value and surprise customers with each delivery, offer a variety of microgreens or seasonal mixes. Define a theme or value for the box, such as "Health Boosting Microgreens" or "Chef's Choice."

- **Determine Subscription Intervals**: Decide weekly, bi-weekly, or monthly subscription intervals based on how often your customers will likely use fresh microgreens.

- **Focus on Quality and Freshness**: Freshness is important with subscription boxes. Plan delivery schedules to ensure each box arrives as fresh as possible, especially if shipping is beyond your local area.

Building Relationships and Maintaining Repeat Customers

Building solid relationships with your first customers is essential for creating a loyal customer base and fostering positive word-of-mouth. Happy, loyal customers will keep returning and spread the word, helping your business grow organically.

Providing Excellent Customer Service

Excellent customer service can set your business apart and turn one-time buyers into repeat customers. Here are some tips:

- **Be Responsive**: Promptly respond to customer inquiries, whether through email, social media, or your website. A quick response shows that you value their interest and are dedicated to a great customer experience.

- **Request Feedback**: After the initial purchase, consider following up with customers to ask for feedback on the product and service. Not only does this show that you care, but it also provides valuable insights into areas where you can improve.

- **Offer Personalized Touches**: Personal touches, like handwritten thank-you notes or occasional discounts, help customers feel appreciated and valued. Offering a small incentive for their next purchase is a great way to encourage repeat business.

Building a Loyalty Program

Loyalty programs are a powerful way to reward repeat customers and build a steady customer base. Here's how to implement one:

- **Points-Based Program**: Offer points for each purchase that can be redeemed for discounts or free products, encouraging ongoing purchases.

- **Discounts for Referrals**: Encourage your customers to refer friends by offering discounts on their next purchase. Referrals help expand your customer base while rewarding loyal customers.

- **Exclusive Perks**: Offer exclusive products or early access to new varieties for loyal customers. Exclusive perks show appreciation and make loyal customers feel special.

Conclusion

Finding and retaining your first customers is a journey that combines strategic outreach, a strong brand presence, and authentic relationship-building. Establishing a loyal customer base goes beyond selling a product; it's about crafting a memorable experience that resonates with your audience, highlights the unique qualities of your microgreens, and reflects the values driving your business. Through targeted efforts—whether it's creating the perfect pitch for chefs and restaurant owners, designing an inviting stand at farmers markets, connecting with specialty food stores, or establishing a seamless online presence—you're reaching diverse customers who share an appreciation for fresh, local, and nutrient-rich greens.

Each customer connection is an opportunity to introduce people to your product's benefits and share the passion and purpose that set your microgreens apart. The genuine care you put into every interaction—answering questions, providing samples, offering helpful suggestions, and following up to ensure satisfaction—demonstrates that your business is more than just a transaction; it's a relationship rooted in quality and trust. Over time, these small but meaningful efforts contribute to a strong foundation of customer loyalty, helping you create a community of repeat buyers who are satisfied and enthusiastic about supporting your brand.

As you continue to grow and engage in customer-focused practices, you'll lay the groundwork for long-term success with customers who return to your business and recommend it to others. Combining quality, consistency, and exceptional service, you cultivate a brand that customers remember and return to. With a loyal base, your microgreens business will thrive, ready to evolve and adapt as you expand your reach, explore new opportunities, and continue building meaningful connections with your customers. This foundation of trust and loyalty will carry your business forward, supporting sustainable growth and making your microgreens a staple in the lives of those you serve.

Chapter 7
Pricing Your Microgreens

Establishing the right price for your microgreens is a delicate yet crucial task that influences your profitability, brand perception, and market competitiveness. Finding this balance requires a thorough understanding of your costs, insights into customer expectations, and knowledge of competitor pricing. An effective pricing strategy communicates the unique value of your product—whether it's the freshness, quality, or sustainable practices you stand behind—while supporting your business goals and appealing to your target customers.

This chapter will delve into the various components contributing to setting a competitive yet profitable price for your microgreens. You'll learn how to calculate costs accurately, assess market demand, and determine a value-driven price point that reflects the quality of your product. Additionally, we'll cover strategies like offering discounts, creating product bundles, and establishing wholesale pricing options to attract new customers, incentivize bulk purchases, and build long-term relationships—all while ensuring that each sale contributes to sustainable growth. By implementing these strategies, you can confidently set prices that resonate with customers and support a thriving microgreens business.

Determining Competitive Yet Profitable Pricing

Setting a profitable and competitive price requires careful analysis of your costs, market positioning, and value proposition. Pricing too high may limit your customer base, while pricing too low may harm your margins and undervalue your product. Here's a step-by-step approach to finding the balance:

Step 1: Understand the Cost Structure

Start by calculating the direct costs associated with producing your microgreens, as these will set the baseline for profitable pricing. Consider all aspects of production:

- **Seeds and Growing Medium**: The price of microgreen seeds varies depending on the variety and supplier, so you must factor in the per-tray cost for both seeds and the growing medium.

- **Water and Nutrients**: Although relatively low-cost, water and added nutrients still contribute to your total expenses. Keep track of the volume used for each crop cycle.

- **Lighting and Energy Costs**: If you grow indoors, account for electricity costs from grow lights or climate control systems. Note your setup's energy consumption and calculate average monthly energy expenses.

- **Labor**: Time is an often overlooked cost. Whether it's your own time or that of hired help, assign a labor cost to each task—seeding, watering, harvesting, packaging, and delivering.

- **Packaging**: Customers expect fresh, hygienic packaging. Calculate the cost per container, label, and any packaging extras, such as biodegradable materials, to reinforce your brand.

- **Transportation**: If you deliver to restaurants, markets, or customers, transportation costs add up quickly. Calculate costs per mile or trip to understand how each delivery impacts your bottom line.

Adding these costs together gives you the total production cost per tray, batch, or weight unit, depending on how you plan to sell your microgreens.

Step 2: Analyze Competitors' Pricing

Researching competitors is essential for determining a price range that aligns with market expectations. Start by gathering data on prices at farmers markets, grocery stores, and online retailers:

- **Farmers Markets**: Visit local farmers markets to assess the prices set by other microgreen vendors. Consider their product quality, packaging, and varieties to see where you fit in.

- **Restaurants and Specialty Stores**: Many restaurants, especially those focused on local or organic ingredients, may have established suppliers with specific pricing. Contact chefs and store managers to understand their purchasing habits and price expectations.

- **Online Retailers**: If you plan to sell online, check prices for microgreens in similar quantities on e-commerce platforms. Many customers pay a premium for the convenience of delivery, which may allow for higher pricing.

By assessing competitors' pricing, you'll gain insight into the market rate and gauge how to position yourself—whether you aim to compete on price, quality, or unique offerings.

Step 3: Identify Your Value Proposition

Your unique value proposition should influence pricing. Growing specialty or hard-to-find varieties, practicing organic growing methods, or using eco-friendly packaging adds perceived value to your product. Highlighting the uniqueness and benefits of your microgreens can justify a higher price point, especially if your product stands out in terms of quality, sustainability, or novelty.

Step 4: Set a Price Range and Adjust as Needed

With cost and competitor analysis, you can set a base price and a target profit margin. Be prepared to adjust your pricing as you refine your processes, negotiate better supply costs, or expand your offerings. Price flexibility can help you remain competitive, adapt to changing demand, and maximize profitability.

Factoring in Costs, Market Demand, and Perceived Value

Balancing your prices with production costs, market demand, and perceived value ensures you stay competitive while maximizing revenue.

Calculating and Covering Costs

For your business to be sustainable, the price of your microgreens must cover both fixed and variable costs while yielding a profit. Fixed costs, such as rent and equipment, remain constant regardless of output, while variable costs, like seeds, water, and labor, fluctuate with production volume. Take into account both types of costs to ensure long-term profitability.

- **Fixed Costs**: If you rent space, purchase equipment, or pay for utilities like water and electricity, divide these costs by your projected sales volume to establish how much each unit must contribute.

- **Variable Costs**: Calculate the per-tray or per-gram costs, including seeds, soil, and labor, to ensure your price can cover these as production scales.

Once costs are covered, the remaining margin can be considered profit, which should align with your business growth and sustainability goals.

Assessing Market Demand

Pricing should also reflect customer demand. High demand may indicate that you can price your products at a premium, while lower demand could suggest a need for competitive pricing. Assessing demand can involve:

- **Market Testing**: Experiment with different price points at farmers' markets to see how customers respond.

- **Customer Feedback**: Ask existing clients for feedback on pricing to gauge their willingness to pay for premium offerings or bulk quantities.

- **Seasonal Trends**: Market demand for microgreens may fluctuate with seasonality and consumer trends, so monitor these factors and adjust pricing as needed to maintain sales.

Enhancing Perceived Value

Perceived value is critical in determining what customers are willing to pay. Highlighting your microgreens' quality, freshness, and nutritional benefits can justify higher prices. Other factors that enhance perceived value include:

- **Premium Packaging**: Using eco-friendly or aesthetically pleasing packaging can make your microgreens feel more premium.

- **Unique Varieties**: If you offer uncommon or hard-to-find varieties, emphasize their distinctive flavors, uses, or health benefits to increase perceived value.

- **Transparency**: Sharing your growing practices, such as organic methods or sustainable approaches, builds trust and enhances value perception among health-conscious customers.

Offering Discounts, Bundles, and Wholesale Pricing

Strategically offering discounts, creating bundles, and setting up wholesale pricing options can help attract new customers, build loyalty, and increase sales volume.

Discounts

Offering occasional discounts can attract new customers and re-engage existing ones. Consider the following types of discounts:

- **First-Time Buyer Discounts**: Offer a small discount for first-time buyers, which can be especially useful online or at markets where customers may be hesitant to try something new.

- **Bulk Purchase Discounts**: For customers purchasing in larger quantities, offer a percentage off or free delivery to encourage bulk buying, which can boost your sales revenue.

- **Seasonal Discounts**: During peak growing seasons, offer discounts on certain varieties to increase sales and manage stock efficiently.

Bundling Products

Product bundles offer value and convenience, helping customers feel they're getting a deal while boosting your average transaction size. Here are some ideas for bundling:

- **Mixed Variety Packs**: Offer a bundle of different microgreen varieties to give customers a sample of your range. This approach can introduce them to new flavors and encourage repeat purchases.

- **Meal Kits**: Partner with local chefs or share recipe cards for meal kits incorporating your microgreens. This adds value for customers who enjoy experimenting with new ingredients.

- **Weekly or Monthly Subscription Bundles**: Create a subscription box that delivers fresh microgreens weekly or monthly. Subscriptions provide consistent revenue and keep your brand top-of-mind with customers.

Wholesale Pricing

If you're working with restaurants, caterers, or specialty grocers, setting up a wholesale pricing structure can help secure larger, consistent orders. Keep these points in mind when establishing wholesale pricing:

- **Volume Discounts**: Offer discounts based on the order volume, rewarding clients who commit to larger quantities with better per-unit pricing.

- **Loyalty Discounts for Long-Term Partnerships**: Consider offering additional discounts or benefits to clients who commit to regular orders, fostering long-term relationships.

- **Customized Orders**: For clients with specific needs, such as unique varieties or exclusive packaging, offer customized options that align with their branding. Customized wholesale packages can justify higher prices while building strong relationships.

Offering discounts, bundles, and wholesale options allows you to meet different customer needs while expanding your reach in the market. Structuring these incentives thoughtfully helps maintain profitability while growing your customer base and encouraging loyalty.

Conclusion

Setting the right price for your microgreens is both an art and a science, requiring a thoughtful balance between covering your costs, aligning with market expectations, and meeting customer needs. A strategic approach to pricing, grounded in a deep understanding of your cost structure and a keen awareness of competitor offerings, allows you to establish a price point that communicates the true value of your product. By assessing and integrating factors like production expenses, customer demand, and perceived value, you can set prices that sustain your business and position it for growth and success.

Introducing versatile pricing options—such as discounts, product bundles, and wholesale rates—enables you to reach a diverse customer base, from individual buyers who value quality and freshness to large-scale clients seeking reliable suppliers. These pricing strategies can enhance customer appeal, encourage repeat purchases, and foster loyalty, allowing you to build strong relationships across multiple sales channels. With a well-rounded pricing approach, your microgreens business will be well-equipped to thrive in a competitive market, adapting to customer preferences and market trends while consistently generating profits.

Ultimately, a well-defined pricing strategy is more than just numbers on a price tag; it reflects your microgreens' quality, care, and unique characteristics. It creates a lasting impression with customers, assures them of your commitment to excellence, and solidifies your reputation in the marketplace. By positioning your prices thoughtfully, you're laying the foundation for a resilient and successful business that meets current demand and is poised for long-term growth and sustainability in the years to come.

Chapter 8
Packaging and Branding

Packaging and branding are essential elements in the success of any microgreens business, as they communicate the quality, personality, and values of your product long before a customer takes their first taste. The right packaging does more than hold your product—it preserves the freshness and flavor of your microgreens, assures customers of their quality, and draws attention in a crowded marketplace. Packaging that balances functionality with an attractive design can make a lasting impression, setting your business apart from competitors and reinforcing the premium nature of your product.

Alongside packaging, clear labeling and a well-defined brand identity build trust, establish authenticity, and fulfill regulatory requirements. Labels communicate vital information, such as nutritional benefits and organic certifications, allowing customers to make informed decisions and assuring them of your commitment to transparency and quality. Meanwhile, a cohesive brand identity ties everything together, reflecting your mission and values in a way that resonates with your target audience, whether they're health-conscious consumers, eco-friendly advocates, or culinary professionals.

This chapter covers designing packaging that balances functionality and aesthetics, exploring methods for maintaining freshness while creating visual appeal, providing guidance on essential labeling requirements—from organic certification to nutritional information—and offering insights into building a brand identity that resonates with target customers, strengthens loyalty, and supports business growth.

Finally, we'll discuss creating a brand identity that reflects your values and connects deeply with your customers, turning one-time buyers into loyal brand advocates. By combining proper packaging, clear labeling, and authentic branding, you'll be well-equipped to establish a memorable and trustworthy presence in the microgreens market.

Packaging That Maintains Freshness and Appeals to Customers

The packaging for microgreens has a dual purpose: it needs to maintain the freshness and quality of the product while attracting customers with an appealing design. Here's how to balance these priorities and create packaging that showcases your microgreens effectively:

Ensuring Freshness and Quality

Microgreens are delicate, requiring packaging that provides optimal protection against moisture loss, bruising, and temperature fluctuations. High-quality packaging preserves flavor, texture, and nutritional value, ensuring a positive customer experience. Here are some key considerations:

- **Breathable Materials**: Microgreens need to "breathe" to prevent excess moisture buildup, which can lead to mold. Look for packaging materials with ventilation holes or use breathable plastic containers to maintain airflow while containing moisture.

- **Durable Containers**: Choose sturdy packaging to prevent crushing during transportation and handling. Rigid plastic or fiber-based clamshells are popular because they protect the product while still being lightweight and easy to stack.

- **Transparency**: Transparent packaging allows customers to see the product's quality, which can boost confidence and encourage purchases. Clear clamshell containers or transparent bags are practical options that let customers inspect the freshness of your microgreens.

- **Environmentally Friendly Options**: Sustainability is a growing concern among consumers. Using recyclable or biodegradable materials can enhance the appeal of your packaging and align with eco-conscious values. Options like compostable clamshells made from cornstarch or recycled paper cartons can help reduce environmental impact, which many customers find appealing.

Designing for Customer Appeal

In addition to functionality, packaging needs to stand out on store shelves or market stalls. An attractive, well-branded package helps communicate the value of your microgreens and makes a memorable impression. Consider the following elements:

- **Color and Design**: Choose colors and design elements that reflect your brand personality. For instance, green hues convey freshness and natural qualities, while clean, minimalist designs can appeal to a premium or health-conscious audience. Including illustrations or images of the microgreens can make the packaging more inviting and informative.

- **Clear Branding**: Your logo, brand colors, and fonts should be consistent across all packaging elements to create a cohesive look. Prominent branding reinforces your identity and helps customers recognize your product at a glance.

- **Convenience Features**: Consider adding resealable closures or easy-to-open tabs for convenience. Resealable packaging helps maintain freshness after opening, making it easier for customers to store and use your product.

- **Information for Customers**: In addition to the required labeling (which we'll cover shortly), consider including usage ideas, storage instructions, or pairing suggestions. These details help customers see your product's versatility and benefits, increasing customer satisfaction and encouraging repeat purchases.

Labeling Laws and Regulations: Organic Certification, Nutritional Information, and Compliance

Labeling isn't just about branding; it's also about providing customers with essential information and complying with regulatory requirements. To build trust with your customers and ensure legal compliance, it's necessary to understand labeling laws related to organic certification, nutritional information, and other packaging guidelines.

Organic Certification

Organic certification can be a valuable selling point if your microgreens are grown organically. However, claiming "organic" on your label involves more than just using organic practices; it requires formal certification from a recognized authority. Here's what to know:

- **Certification Process**: To label your product as organic, you must go through a certification process with an accredited agency, such as the USDA in the United States or another recognized organization in your country. This process typically involves on-site inspections, records of organic practices, and annual renewals.

- **Labeling Requirements**: Once certified, you can use the official organic seal on your packaging, which adds credibility and signals to customers that your microgreens meet organic standards. Be mindful that "organic" claims are regulated, and misuse can result in fines or loss of certification.

- **Marketing Benefits**: Organic labeling appeals to health-conscious and eco-minded customers willing to pay a premium for certified organic products. Displaying the organic seal prominently on your packaging can make your product more attractive to this growing customer base.

Nutritional Information

Nutritional labeling offers customers insights into the health benefits of your microgreens. Depending on your region, nutritional labels may be required, so check local regulations. Here's what to consider:

- **Standard Nutritional Labels**: If required, standard nutritional labels typically include information like calorie count, macronutrients (fats, proteins, and carbohydrates), and vitamin content. Microgreens are often rich in vitamins, minerals, and antioxidants, so including these details can highlight their health benefits.

- **Highlighting Key Nutrients**: Even if a full nutritional label isn't mandatory, consider highlighting important nutrients, such as Vitamins A, C, and K, potassium, or fiber, on your packaging. Displaying nutritional benefits can attract health-focused customers who prioritize nutrient-dense foods.

- **Transparency and Trust**: Transparent labeling helps build trust with customers, reassuring them of your commitment to quality and honesty. These additional details can distinguish your product from competitors if you go above and beyond with organic practices, eco-friendly methods, or special growing techniques.

General Labeling Compliance

To avoid penalties and ensure your microgreens are compliant with local food labeling laws, be aware of general requirements, which may include:

- **Ingredients List**: Some jurisdictions require an ingredient label, even for single-ingredient products like microgreens. List "microgreens" and specify the type if it's a blend.

- **Expiration or Harvest Dates**: Fresh produce often benefits from a "packed on" or "best by" date to assure customers of its freshness. Clear date labeling is especially important for microgreens, as freshness impacts quality.

- **Country of Origin**: Some countries require a label indicating where the product was grown. This can be a positive feature if you're marketing locally-grown produce, as it can attract customers who value local food sources.

Labeling compliance ensures that your business avoids legal risks, meets industry standards, and builds consumer trust by presenting accurate, honest product information.

Creating a Brand Identity That Resonates with Your Target Audience

A strong brand identity does more than make your product recognizable; it creates a lasting connection with customers by conveying your values, mission, and unique qualities. A brand identity that resonates with your target audience drives sales and fosters loyalty, encouraging customers to choose your microgreens over others. Here's how to build a brand that speaks to your ideal customers:

Defining Your Brand Values and Mission

Start by clearly defining your brand values and mission, as these will guide your messaging, visuals, and customer interactions. Reflect on what sets your microgreens apart and why you started this business. Some values to consider include:

- **Sustainability**: If you use eco-friendly practices, emphasize your commitment to sustainability. This can be particularly appealing to environmentally conscious consumers.

- **Health and Wellness**: Microgreens are known for their nutrient density, so positioning your brand as a health-oriented choice can attract customers focused on wellness.

- **Community and Local Support**: Highlighting a commitment to supporting local agriculture or community initiatives can resonate with customers who value local business and ethical sourcing.

Your brand values should be reflected in everything you do, from packaging to customer service, creating a cohesive identity that is easy for customers to understand and connect with.

Crafting a Visual Identity

Your brand's visual elements—logo, colors, fonts, and imagery—play a major role in how customers perceive your product. Aim for a cohesive and memorable visual identity that reflects your values and appeals to your target audience:

- **Logo Design**: A well-designed logo is a key branding element. Whether you work with a professional designer or create a simple logo, ensure it's versatile and recognizable across all platforms.

- **Color Palette**: Colors evoke specific emotions and associations. For example, green can signify health and freshness, while earthy tones evoke natural and organic qualities. Choose colors that align with your brand's personality and values.

- **Fonts and Typography**: The fonts you use on packaging, website, and marketing materials contribute to your brand's tone. Modern, clean fonts can suggest sophistication and quality, while playful, organic fonts might appeal to a more casual, eco-friendly audience.

Connecting with Your Audience Through Messaging

Your brand messaging should speak directly to your target audience, addressing their needs, interests, and values. Consider how you present your product's benefits and how your story can make your brand memorable. For example:

- **Highlighting Product Benefits**: Microgreens offer numerous health benefits, so make sure your messaging communicates these advantages. Phrases like "nutrient-packed," "freshly harvested," and "locally grown" can resonate with health-conscious customers.

- **Telling Your Story**: Sharing the story behind your brand—such as your passion for sustainable farming or journey into microgreen cultivation—adds a personal touch. Customers often connect emotionally with brands that share their values or have inspiring stories.

- **Creating Consistent Messaging**: Use consistent language and tone across all customer touchpoints, from social media posts to in-person interactions. Consistent messaging reinforces your brand identity, making it more memorable for customers.

A well-defined brand identity, supported by thoughtful packaging and labeling, helps position your microgreens business for success. When branding resonates with customers, it builds loyalty, encourages word-of-mouth referrals, and fosters a lasting connection with your audience.

Conclusion

Effective packaging and a compelling brand identity are crucial to establishing a successful microgreens business. Thoughtful packaging preserves freshness and quality and engages customers, creating a memorable first impression that sets your product apart from competitors. Compliance with labeling regulations further strengthens customer trust, providing transparency and reassuring them of your commitment to quality and authenticity. A brand identity that authentically reflects your values and speaks to your target audience builds connection and loyalty, fostering a positive perception beyond the product.

When customers recognize and trust your brand, they're far more likely to become repeat buyers, contributing to a loyal customer base that supports long-term growth. With a strategic approach to packaging, branding, and labeling, your microgreens can stand out in any market setting—whether on store shelves, at farmers markets, or online—attracting customers who value quality, freshness, and the principles your brand represents. Establishing this foundation paves the way for sustainable success, giving your microgreens a powerful presence in a competitive landscape.

Chapter 9
Expanding Your Product Line

Expanding your product line is a powerful way to unlock new growth opportunities, attract a wider range of customers, and ultimately increase sales. In the microgreens business, a diverse product offering allows you to stand out in a competitive market and cater to various customer needs and preferences. Expanding can take many forms, from introducing new microgreen and sprout varieties to developing value-added products that provide customers with ready-to-use options. Additionally, strategic partnerships with local businesses such as restaurants, specialty grocers, and wellness centers can deepen your community presence and broaden your customer base.

By diversifying your product line, you create multiple revenue streams, making your business more resilient and adaptable to changing market demands. New varieties keep your offerings fresh and interesting, while value-added products such as pre-made salads or DIY grow kits cater to the growing demand for convenience and versatility. Collaborating with local businesses enhances brand visibility and builds trust within your community, allowing you to reach more customers and establish your brand as a reliable supplier.

This chapter will guide you through identifying high-demand microgreen varieties, exploring opportunities for value-added products, and forming mutually beneficial partnerships with local businesses. These strategies are designed to help your business expand sustainably, align with customer interests, and strengthen its market position for long-term growth and success.

Adding New Varieties of Microgreens and Sprouts

Introducing new varieties of microgreens and sprouts is one of the simplest ways to expand your product line and keep your offerings fresh and exciting. Different varieties appeal to different tastes, culinary uses, and nutritional preferences, allowing you to reach a broader audience. Growing new types of microgreens can differentiate your business from competitors who may offer only the most common varieties. Here's a guide to selecting and growing new varieties:

Identifying High-Demand Microgreens and Sprouts

To choose the best varieties to add, start by identifying those in high demand among your target audience. For example:

- **Popular Microgreens**: Many customers are familiar with popular microgreens such as pea shoots, sunflower shoots, and radish greens. However, offering unique varieties like red amaranth, cilantro microgreens, or mustard greens can attract adventurous eaters, chefs, and health-conscious customers looking for variety.

- **Nutrient-Dense Sprouts**: Adding nutrient-dense options like broccoli, chervil, and alfalfa sprouts can cater to customers focused on health and wellness. These sprouts are easy to grow and are recognized for their high levels of vitamins and antioxidants, appealing to health-focused consumers.

- **Culinary Uses**: Research culinary trends to identify microgreens commonly used by chefs. For example, spicy radish or delicate basil microgreens are popular among chefs looking to enhance their dishes. Offering specialty varieties for culinary use can help you connect with restaurant clients interested in creative ingredients.

Experimenting with Flavor Profiles and Colors

Offering a diverse range of flavors and colors enhances the appeal of your product line. Experimenting with different types can help you create a visually stunning and flavorful selection, such as:

- **Bold Flavors**: Varieties like arugula, mustard, and radish microgreens bring bold, peppery flavors that add a punch to dishes. Highlight these options as flavor enhancers for salads, sandwiches, or garnishes.

- **Colorful Options**: Varieties like red cabbage, beet, and red amaranth microgreens add striking colors that make dishes more visually appealing. Emphasize the aesthetic value of colorful microgreens for customers interested in presentation, including chefs and caterers.

- **Milder Options for Versatility**: For customers who prefer subtle flavors, consider varieties like pea shoots, sunflower greens, or mild basil microgreens. These options provide versatility, working well as bases for salads or as a fresh, crunchy addition to any dish.

Testing New Varieties

When adding new varieties, start with a small batch to gauge demand and ensure your growing setup is suitable. Here's a process to guide you:

1. **Grow a Sample Batch**: Begin by growing a limited amount of the new variety. This will allow you to assess the growing requirements and quality of the final product.

2. **Gather Customer Feedback**: Offer samples to customers at markets or include them in chef consultations. Feedback from these early trials can help you decide if the new variety should become a regular offering.

3. **Scale Production Based on Demand**: If the feedback is positive and demand is consistent, gradually increase production to meet customer interest.

Introducing new varieties allows you to expand your product line with minimal risk and effectively respond to customer preferences.

Offering Value-Added Products

In addition to fresh microgreens, value-added products can increase the appeal of your business and encourage repeat purchases. Value-added products, such as microgreen salads and grow kits, offer convenience and versatility, allowing customers to enjoy microgreens in new ways. Here are several types of value-added products that can complement your existing offerings:

Ready-to-Eat Microgreen Salads

Microgreen salads are a convenient, healthy option for customers looking for fresh, ready-to-eat products. Preparing pre-mixed salads makes it easy for customers to incorporate microgreens into their diets. Here's how to develop and market these salads:

- **Mixing Varieties for Flavor and Nutrition**: Create salad mixes by combining several microgreens to achieve a balanced flavor profile and nutrient content. For example, a mix of sunflower, arugula, and radish microgreens provides a combination of mild, spicy, and nutty flavors and a wide range of vitamins.

- **Adding Complementary Ingredients**: Enhance the appeal of your salads by including complementary ingredients such as edible flowers, herbs, or seasonal fruits and vegetables. For instance, adding fresh berries or citrus segments to a microgreen salad mix can create a flavorful, eye-catching product.

- **Packaging for Freshness and Convenience**: Use attractive, resealable packaging that keeps salads fresh and easy to transport. Consider offering single-serving sizes for grab-and-go convenience as well as larger sizes for families or meal prep.

Microgreen-Based Juices and Smoothies

Juices and smoothies made from microgreens offer customers a unique way to enjoy their nutritional benefits. Microgreen-based beverages can cater to customers interested in health, wellness, and functional foods:

- **Blending Nutrient-Dense Greens**: Experiment with nutrient-dense greens like kale, broccoli, and wheatgrass microgreens to create energy-boosting drinks. These microgreens provide concentrated nutrients that appeal to health-conscious consumers.

- **Combining with Fruits and Superfoods**: Create unique flavor profiles by combining microgreens with fruits, like apples or berries, and superfoods like chia seeds or ginger. Highlight the health benefits and refreshing taste of each blend.

- **Offering Seasonal Flavors**: Introduce limited-edition flavors based on seasonal produce, which can encourage customers to try new varieties. Seasonal offerings keep your product line dynamic and give customers a reason to return for new options.

Microgreen Grow Kits

Grow kits empower customers to cultivate their own microgreens at home, providing an educational experience and a lasting connection to your brand. Grow kits are ideal for customers who value DIY projects or wish to experience fresh greens on demand:

- **Including All Necessary Supplies**: A basic grow kit should include seeds, a growing medium, trays, and detailed instructions. Making it easy for beginners to get started growing their microgreens.

- **Customization Options**: Offer kits with different microgreen varieties so customers can choose based on taste or nutritional preference. For example, create beginner kits with easy-to-grow varieties like radish and pea shoots and specialty kits with unique seeds for more advanced growers.

- **Encouraging Repeat Sales**: Once they have a grow kit, customers will likely need refills. Offer seed refills or growing medium replenishments, allowing customers to continue their home-growing journey conveniently.

Collaborating with Local Businesses

Collaborating with local businesses, such as restaurants, grocery stores, and cafes, can expand your customer base and increase brand exposure. By establishing strategic partnerships, you can reach new markets and develop a reputation as a reliable, high-quality supplier in your community. Here's how to approach collaborations effectively:

Supplying Microgreens to Restaurants and Caterers

Restaurants and caterers often seek fresh, unique ingredients to elevate their dishes, making them ideal partners for microgreens producers. Building relationships with local chefs can provide consistent orders and valuable market insights. Here's how to establish these connections:

- **Understand the Chef's Needs**: Chefs prioritize quality, consistency, and freshness. Reach out to understand their specific needs, preferred varieties, and seasonal requirements. Specialty varieties, like edible flowers or vibrant microgreens, can make your product more appealing to culinary professionals.

- **Offer Customization Options**: Some chefs may want microgreens in specific sizes or mixes for particular dishes. Be open to customizing orders or creating unique blends for regular clients to strengthen relationships and secure long-term business.

- **Provide Samples and Tasting Opportunities**: Sampling is an effective way to showcase your product's quality. Arrange a tasting session with chefs or offer complimentary samples, allowing them to experience the flavor and presentation of your microgreens.

Selling Microgreens at Local Grocery Stores and Markets

Local grocery stores and specialty markets are excellent venues for selling microgreens. They provide access to health-conscious customers and culinary enthusiasts. Establishing a presence in these retail locations can increase visibility and make it easy for customers to purchase your product.

- **Highlight Freshness and Local Origins**: Many shoppers prioritize fresh, local products, so emphasize that your microgreens are locally grown. Display labels like "Locally Grown" or "Farm Fresh" on your packaging to reinforce this message.

- **Ensure Shelf Appeal**: Retail packaging should be eye-catching and informative. Use transparent packaging to show the product's freshness and include information on flavor profiles, nutrition, and suggested uses.

- **Maintain Consistent Supply**: Reliable supply is crucial when working with retail partners. Develop a production schedule that aligns with store inventory needs, ensuring they consistently supply your microgreens to build customer loyalty.

Collaborating with Health and Wellness Businesses

Health and wellness businesses, such as juice bars, fitness centers, and wellness cafes, often cater to customers interested in fresh, nutrient-dense foods. Collaborating with these businesses can help position microgreens as a strategic component of a healthy lifestyle.

- **Co-Branding Opportunities**: Partner with wellness businesses to co-brand products like smoothies, salads, or wellness bowls featuring your microgreens. Co-branded products can create a mutually beneficial arrangement that promotes both businesses.

- **Participate in Wellness Events**: Many wellness businesses host events, such as health workshops or fitness classes, that attract an engaged, health-focused audience. Offering samples or selling microgreens at these events can introduce your product to a new group of potential customers.

- **Educational Content**: Wellness businesses often prioritize customer education. Provide informational material on the health benefits of microgreens that they can share with clients, helping you establish a reputation as a knowledgeable and trusted brand.

Conclusion

Expanding your product line through new microgreen varieties, value-added products, and strategic partnerships can propel your microgreens business toward sustainable growth, increased visibility, and a broader customer base. By introducing unique varieties and experimenting with distinct flavors, you attract health-conscious customers seeking diversity, vibrant aesthetics, and nutritional benefits. Value-added offerings, such as ready-to-eat salads, juices, or DIY grow kits, meet the growing demand for convenience and interactive experiences, helping you appeal to everyday consumers and culinary enthusiasts. Meanwhile, collaborations with local businesses like restaurants, specialty stores, and wellness centers establish your brand as a trusted, local supplier, strengthening community ties and enhancing brand awareness.

A well-diversified product line amplifies revenue potential and reinforces your business's adaptability and relevance in a competitive market. Thoughtfully expanding your offerings builds a brand identity that resonates with customers' evolving needs, demonstrating your commitment to quality and innovation. As your business grows, your expanded product line will offer customers new and exciting ways to enjoy microgreens' freshness, flavor, and versatility, fostering loyalty and positioning your business for long-term success. Through this strategy of thoughtful expansion, your microgreens business can cultivate a lasting impact and continue thriving in the marketplace.

Chapter 10
Marketing and Promoting Your Business

Effective marketing and promotion are vital for growing a thriving microgreens business. In today's competitive marketplace, having a quality product is just the beginning. To truly stand out, it's essential to actively engage with your audience, communicate the unique benefits of your microgreens, and build strong brand awareness that keeps customers coming back. A well-crafted marketing strategy helps you reach a wider customer base and strengthens your brand identity, turning first-time buyers into loyal advocates who support your business over the long term.

Strategic marketing provides opportunities to showcase your microgreens' health benefits, freshness, and sustainability, connecting with customers on a deeper level. By leveraging digital channels like social media, participating in local events, and tapping into growing trends in health and wellness, you can create a recognizable and trustworthy brand. In this chapter, we'll explore how to use social media and online marketing to broaden your reach, implement local advertising strategies that resonate with the community, and incorporate wellness-focused messaging to attract health-conscious customers. Additionally, we'll discuss the power of building a brand story centered on sustainability and well-being—two values that resonate deeply with modern consumers. These combined efforts will expand your customer base and solidify your business as a respected and reliable source of quality microgreens.

Using Social Media and Online Marketing to Grow Your Reach

Social media and online marketing offer dynamic platforms for expanding your reach, engaging directly with customers, and cultivating a loyal community around your brand. By implementing creative, targeted online strategies, you can effectively highlight the benefits of your microgreens, educate your audience, and establish a solid digital presence that aligns with your brand's values. These tools allow you to connect with a wider audience and create lasting impressions that build trust and loyalty. Here's how to maximize the potential of social media and other online channels:

Choosing the Right Platforms

Each social media platform offers unique opportunities to connect with different segments of your target audience, so choosing the right ones can help you maximize engagement and effectively showcase your microgreens. Here's a guide to using the platforms that are best suited for promoting your brand:

- **Instagram**: With its strong focus on visuals, Instagram is ideal for highlighting the vibrancy, freshness, and appeal of your microgreens. Features like stories, reels, and carousel posts allow you to share recipes, growing tips, and behind-the-scenes glimpses into your production process, creating a personal connection with your followers. High-quality photos and short videos can showcase different varieties and seasonal offerings and visually capture the beauty of your products. Engaging captions and interactive features like polls and Q&As can further boost engagement and help build a loyal community.

- **Facebook**: Known for its community-oriented approach, Facebook is an excellent platform for connecting with local customers and joining groups related to health, wellness, and sustainable food. Facebook Events can be used to promote appearances at farmers markets, workshops, or special promotions, while targeted ads allow you to reach people interested in organic food, local produce, and health-conscious living. Facebook's wide-ranging audience also allows you to target specific demographics, such as families or wellness-focused consumers, making it a versatile tool for expanding your reach.

- **YouTube**: YouTube is ideal for sharing longer, informative videos that establish your brand as an authority in the microgreens industry. Create content such as tutorials for growing microgreens at home, interviews with chefs who use your products, or educational videos about the nutritional benefits of various microgreen varieties. This type of content appeals to viewers interested in health, gardening, and DIY projects and can help you attract a dedicated following that values in-depth knowledge and expertise.

- **Pinterest**: As a visually-driven platform that excels in DIY, food, and health-related content, Pinterest is perfect for sharing recipes, meal inspiration, and creative ways to use microgreens. By pinning high-quality images of your products and linking them back to your website, you can drive traffic, boost brand awareness, and reach an audience that actively searches for health-conscious and visually appealing ideas. Pinterest also allows you to create boards dedicated to specific themes, such as "Healthy Recipes" or "Microgreen Growing Tips," making it easy for users to discover your content.

- **TikTok**: With its rapid growth and popularity among younger audiences, TikTok provides a unique opportunity to share quick, engaging, and creative content that can go viral. Microgreens fit well into popular TikTok niches like health, wellness, and sustainable food. Use short, visually captivating videos to demonstrate easy microgreen recipes, creative plating ideas, or the process of growing microgreens from start to finish. TikTok's algorithm favors engaging content, so experiment with trending sounds, challenges, or hashtags related to health and wellness. TikTok is also a platform where authenticity is highly valued, so showing the day-to-day reality of your business—whether it's harvesting, packaging, or delivering to local restaurants—can help humanize your brand and connect with viewers on a personal level.

Choosing platforms that align with your brand's strengths and target audience can amplify your reach, engage your customers, and create a solid online presence showcasing your microgreens' unique qualities. Each platform offers its strengths, so a well-rounded strategy using a few key channels will help you maximize engagement and build a lasting community around your brand.

Content Ideas to Engage Followers

The content you share online is your brand's voice, so aim to make it engaging, informative, and aligned with your brand values. Here are some ideas to keep your audience interested:

- **Recipe Sharing**: Regularly post recipes that incorporate your microgreens, such as salads, sandwiches, or smoothie bowls. Sharing easy, appealing recipes inspires followers to use your product more frequently.

- **Health and Nutrition Tips**: Share posts about the health benefits of different microgreens, explaining the nutrients they offer and how they support wellness. These posts educate your audience and establish your brand as a knowledgeable resource for healthy living.

- **Behind-the-Scenes Content**: Show snippets of your growing process, market setups, or even team members at work. This type of content builds transparency and trust, giving followers a closer look at the people and processes behind your brand.

- **User-Generated Content**: Encourage customers to share photos of how they use your microgreens and tag your account. Reposting user-generated content not only fosters community but also provides authentic testimonials for potential customers.

Paid Advertising and Online Marketing

In addition to organic content, paid online advertising can help amplify your reach and attract new customers:

- **Facebook and Instagram Ads**: Both platforms allow you to create highly targeted ad campaigns based on location, interests, and demographics. For example, target people interested in healthy eating, farmers markets, or organic food.

- **Google Ads**: If you have an e-commerce site, using Google Ads can drive traffic from people actively searching for microgreens or similar products. Consider using search ads or display ads to promote your business.

- **Influencer Collaborations**: Partner with local influencers, chefs, or health advocates who can promote your microgreens to their followers. Influencers offer authentic promotion and can introduce your product to new audiences who trust their recommendations.

Local Advertising Strategies: Community Events and Partnerships

While online marketing is essential for reaching a broad audience, local advertising, and community involvement are invaluable for building trust and recognition within your area. Promoting your brand through community events, partnerships with local businesses, and word-of-mouth referrals fosters a loyal customer base. It establishes your business as a trusted staple in the community. This hometown connection boosts credibility and strengthens relationships with customers who value supporting local enterprises.

Participating in Community Events

Getting involved in local events is a great way to increase brand visibility and connect with your community face-to-face:

- **Farmers Markets**: Setting up a stand at local farmers markets allows you to interact directly with customers, share product samples, and educate them on the benefits of microgreens. Markets provide a platform to build personal relationships and make your business a familiar name within the community.

- **Health and Wellness Fairs**: Attend local health fairs or wellness events to showcase your microgreens to a health-focused audience. Offer samples, share educational materials, and engage with attendees about the nutritional benefits of your products.

- **Cooking Classes or Demos**: Partner with local chefs or wellness coaches to host cooking classes or demonstrations using your microgreens. These events allow you to showcase your product in creative, practical ways and provide valuable exposure for your brand.

Building Local Partnerships

Collaborating with local businesses can mutually benefit both parties, increase brand exposure, and strengthen your business's community ties:

- **Restaurants and Cafés**: Establish partnerships with restaurants or cafés that prioritize fresh, local ingredients. Offer to supply them with unique varieties of microgreens and ask if they can feature your brand on their menu. This arrangement helps position your brand as a premium local option and introduces your product to new customers.

- **Grocery Stores and Health Food Shops**: Stocking your microgreens in local grocery stores or health food shops provides access to customers who value fresh, organic produce. Negotiate display arrangements to ensure your product is presented prominently, ideally near other healthy, fresh foods.

- **Yoga Studios and Fitness Centers**: Many wellness spaces look for ways to enhance their clients' health journeys. Offer collaborations with local yoga studios or gyms, where you can share informational brochures or samples or even co-host wellness events.

Tapping into the Health and Wellness Trend in Marketing

With their rich nutrient profile and versatile uses, microgreens are ideally suited for the health and wellness market. Positioning your product as a wellness-focused, nutrient-packed choice can attract health-conscious customers who prioritize fresh, natural foods. Here's how to effectively align your brand with current health and wellness trends:

Highlighting Nutritional Benefits

Use educational marketing to explain the nutritional benefits of microgreens. Different microgreens offer various health advantages, so focus on key nutrients in each variety:

- **Vitamins and Antioxidants**: Emphasize that microgreens are packed with essential vitamins like A, C, and K, as well as antioxidants that support immune health and overall wellness.

- **Unique Nutritional Profiles**: Certain microgreens, like broccoli or kale, are known for specific benefits, such as aiding in detoxification or supporting heart health. Including this information on social media or packaging can help customers choose varieties that meet their health goals.

- **Functional Benefits**: Highlight functional benefits, such as microgreens that support energy, digestion, or skin health. For example, pea shoots are known for their high protein content, making them appealing to customers who focus on plant-based protein sources.

Educating Customers on Integrating Microgreens into Their Diet

Customers may be unfamiliar with how to use microgreens, so providing easy, practical ways to incorporate them into meals can encourage more frequent purchases:

- **Recipe Cards or Booklets**: Include recipe cards or a small booklet with purchase, giving customers ideas for meals, snacks, and drinks that incorporate microgreens.

- **Social Media Tips and Challenges**: Run social media campaigns that offer tips for integrating microgreens into daily diets or challenge followers to try a new microgreen-based recipe each week.

Collaborating with Health Influencers

Health and wellness influencers can amplify your message by sharing your microgreens with their followers. Partnering with nutritionists, fitness coaches, or health bloggers who align with your brand values can give your product a credible endorsement, reaching potential customers who are highly engaged in the health and wellness space.

Building a Brand Story Around Sustainability and Health

Your brand story is a powerful tool for connecting with customers on a deeper level. When built around sustainability, health, and local impact, your brand story can resonate with conscious consumers who prioritize these values.

Emphasizing Sustainable Practices

If sustainability is a core value of your business, highlight the eco-friendly aspects of your growing practices:

- **Environmentally Friendly Growing**: Explain any sustainable practices in your production process, such as using organic methods, minimizing water usage, or relying on renewable energy. Sharing this story on your website or social media reinforces your commitment to sustainability.

- **Eco-Friendly Packaging**: If your packaging is recyclable or compostable, make this clear on labels and marketing materials. Many customers appreciate brands that try to reduce environmental impact, which can be a deciding factor for them.

Telling a Personal Brand Story

If your business has a unique origin story, share it with your audience to create an authentic connection:

- **Why You Started**: Share the inspiration behind your business, whether it's a passion for health, a desire to provide quality local produce or a commitment to sustainable agriculture.

- **Behind-the-Scenes Content**: Offer glimpses into your daily operations, like planting or harvesting, to show the care and dedication that goes into each batch of microgreens. These personal touches give your brand character and authenticity.

Positioning Your Brand as a Health and Wellness Leader

As you establish a strong brand story, position your business as a trusted resource for health-conscious customers:

- **Educational Content**: Consistently sharing valuable information—such as the benefits of microgreens, recipes, or wellness tips—builds your reputation as a knowledgeable brand in health and wellness.

- **Community Engagement**: Actively participate in wellness communities, whether through partnerships, events, or online engagement. Becoming part of a broader health and wellness network enhances your brand's credibility and reach.

Conclusion

Effectively marketing and promoting your microgreens business can drive meaningful growth, helping you reach a broad audience, cultivate a loyal customer base, and build a powerful, lasting brand presence. You can engage diverse groups through strategic social media marketing, local advertising efforts, and collaborations with like-minded businesses—from local shoppers and culinary enthusiasts to health-conscious consumers who value nutrient-dense foods. By aligning your brand with the health and wellness movement and crafting a story rooted in sustainability and fresh, local goodness, you position your microgreens as not just a product but a lifestyle choice that resonates with today's values.

This thoughtful, multi-channel approach to marketing reinforces your brand's unique identity, helping it become a recognized and trusted name in the microgreens industry. With a blend of creativity, consistency, and strategic messaging, your brand can thrive in a competitive market, attracting customers who appreciate the quality and ethos behind every product. As you continue to grow, your marketing efforts will pave the way for long-term success, making your microgreens business a go-to source for those who value health, sustainability, and community connection.

Chapter 11
Navigating Challenges And Competition

Running a successful microgreens business brings unique challenges that require careful planning, adaptability, and forward-thinking strategies. Each aspect demands a proactive approach, from managing unpredictable supply chains to handling crop failures and navigating an increasingly competitive landscape. As the popularity of microgreens continues to grow, so does the competition, making it essential to overcome these challenges, stand out, and stay ahead in a dynamic market.

With a well-rounded strategy, you can turn potential obstacles into opportunities for growth and differentiation. In this chapter, we'll delve into practical strategies for managing supply chain disruptions and minimizing risks from crop failures. We'll also cover approaches for handling market competition, from understanding competitor offerings to differentiating your brand in ways that resonate with customers. Finally, we'll explore how to foster innovation and remain attuned to market trends, positioning your business as a leader in the industry. These combined efforts will support your business's resilience and ensure its lasting success as a trusted source of quality microgreens.

Managing Supply Chain Issues and Crop Failures

A reliable supply chain and healthy crop yields are the foundation of any microgreens business. However, both can be affected by unexpected disruptions, from supplier shortages to crop diseases and equipment malfunctions. To build a resilient business, it's essential to have contingency plans and strategies that help you manage these challenges effectively.

Diversifying Suppliers and Managing Inventory

Relying on a single supplier for seeds, growing media, or packaging materials can lead to disruptions if they experience shortages or delays. Diversifying your suppliers allows you to maintain steady access to the essentials, even if one supplier experiences an issue. Here's how to manage supplier relationships and inventory effectively:

- **Build Relationships with Multiple Suppliers**: Establish connections with at least two suppliers for each critical resource, such as seeds and packaging. Look for reliable, high-quality sources that can offer backup options if your primary supplier cannot fulfill an order.

- **Create Inventory Buffers**: Maintain a buffer of essential supplies like seeds, soil, and packaging. This reserve can keep your business running during short-term disruptions, giving you time to source new suppliers if needed.

- **Use Local Suppliers When Possible**: Sourcing locally reduces shipping delays and makes building strong relationships with suppliers investing in local businesses easier. Local sourcing can also appeal to customers who value sustainability.

Handling Crop Failures and Preventing Losses

Crop failures can occur for various reasons, including disease, poor-quality seeds, and environmental factors. Although not all crop failures can be prevented, there are measures you can take to minimize risk and recover quickly if issues arise:

- **Regular Crop Rotation**: Rotating your crop types helps prevent diseases and pests from becoming established. For example, alternate between fast-growing microgreens like radish and slower-growing varieties like basil to reduce pest and disease cycles.

- **Monitor Growing Conditions**: Invest in tools that measure moisture, temperature, and light conditions to optimize the environment. Small changes in temperature or humidity can impact yields, so close monitoring allows you to catch issues early and make adjustments.

- **Have Backup Seeds and Varieties**: Keeping a stock of backup seeds will enable you to plant new batches quickly if a crop fails. If you're experimenting with new varieties, ensure you have established crops growing simultaneously, reducing the risk of losing an entire cycle.

- **Experiment with Controlled Environments**: If feasible, consider investing in controlled-environment agriculture (CEA) setups such as grow tents, greenhouses, or climate-controlled rooms. CEA can help you manage growing conditions year-round, reducing susceptibility to external environmental factors.

Dealing with Competition and Market Saturation

As the popularity of microgreens continues to rise, competition within the market is intensifying. With more producers entering the field, market saturation can become a challenge, making it essential to differentiate your brand and offerings. A clear strategy for dealing with competition allows your business to maintain its unique appeal and attract loyal customers.

Understanding Your Competitors and Their Offerings

Knowing who your competitors are and what they offer provides insights into how you can set yourself apart. Conducting a competitor analysis is an effective way to evaluate the market and pinpoint opportunities to differentiate:

- **Identify Key Competitors**: Identify direct and indirect competitors in your local area and online. Direct competitors may be other microgreens businesses, while indirect competitors could be grocery stores or organic produce suppliers.

- **Analyze Their Strengths and Weaknesses**: Examine competitors' products, pricing, packaging, and customer engagement. Identify gaps in their offerings that you can fill—whether it's a unique product, better packaging, or a stronger focus on sustainability.

- **Monitor Pricing Strategies**: Evaluate competitors' pricing strategies to understand what customers expect to pay for microgreens. Aim to set competitive prices while ensuring your value is clear, especially if you offer premium, organic, or locally-grown varieties.

Differentiating Your Product and Brand

Differentiating your brand from competitors helps attract customers looking for something unique. You can stand out in a saturated market by offering distinct products or emphasizing certain qualities. Here are some ideas for differentiation:

- **Offer Unique Varieties**: Introducing less common varieties, such as amaranth, nasturtium, or shiso, can attract customers interested in experimenting with new flavors or garnishing options. Highlight each unique variety's flavor profiles, nutritional benefits, and culinary uses.

- **Emphasize Quality and Freshness**: Freshness is crucial in microgreens, so emphasize short harvest-to-shelf times. Consider offering same-day or next-day delivery to restaurants and local customers to ensure they receive the freshest product possible.

- **Focus on Sustainability**: Many customers prioritize eco-friendly options, so communicate your commitment to sustainable practices. This could include using recyclable packaging, sourcing locally, or adopting energy-efficient growing methods.

- **Create Custom Blends**: Custom blends tailored to culinary use or health benefits can add a level of personalization that differentiates your product. For example, create a "detox" blend of microgreens rich in antioxidants or a "salad blend" with mild, versatile flavors.

Building Strong Relationships with Customers

In a competitive market, customer loyalty can be a major advantage. Building relationships with your customers can lead to repeat business and positive word-of-mouth, helping you maintain a steady client base even as competition increases:

- **Personalized Customer Service**: Providing responsive and personalized customer service builds trust and strengthens relationships. For example, send follow-up emails to check customer satisfaction or offer personalized product recommendations.

- **Reward Loyalty**: Implement a loyalty program or offer discounts to returning customers. Loyalty programs encourage repeat business and create a sense of appreciation among long-term customers.

- **Gather Customer Feedback**: Regularly ask for feedback on your products, packaging, and service. Customers appreciate being heard, and feedback can offer valuable insights into what's working and where you can improve.

Staying Innovative and Ahead of Market Trends

Innovation and trend awareness are essential for staying relevant in a rapidly evolving market. By staying ahead of trends, you can anticipate customer preferences, introduce products that align with evolving demands, and ensure your business remains competitive and adaptable.

Monitoring Industry Trends

Keeping a pulse on industry trends helps you anticipate changes in customer preferences and adjust your offerings accordingly. Here's how to stay updated:

- **Follow Industry Publications and Forums**: Subscribe to publications and online forums that focus on microgreens, sustainable agriculture, or health food trends. Platforms like LinkedIn groups, Reddit, and specialized websites often feature discussions on the latest industry developments.

- **Attend Trade Shows and Conferences**: Industry events provide valuable insights into emerging trends, new growing techniques, and networking opportunities with other professionals. Attending these events lets you discover the latest products, technologies, and practices that could benefit your business.

- **Engage with Customers on Social Media**: Social media is an excellent way to keep up with customer preferences and spot trends early. Pay attention to posts, comments, and messages to understand what customers are excited about and what they're looking for in their food choices.

Experimenting with New Products and Services

Regularly experimenting with new products and services allows you to expand your offerings and remain competitive. By introducing innovative products, you can cater to evolving market demands and appeal to a broader audience:

- **Value-Added Products**: Expand beyond fresh microgreens with value-added products, such as microgreen-based juices, salads, or DIY grow kits. Value-added products offer convenience and appeal to customers seeking quick, healthy options.

- **Seasonal and Limited-Edition Offerings**: Creating limited-edition or seasonal products can add excitement and urgency to your offerings. For example, offer seasonal microgreens or unique blends that align with popular trends or health goals, like "immune-boosting" or "detox" blends.

- **Educational Content and Workshops**: Many customers are interested in learning more about microgreens, nutrition, and sustainable agriculture. Consider hosting workshops, webinars, or online classes to engage customers and establish your brand as an industry expert.

Leveraging Technology and Automation

Investing in technology and automation can streamline operations and enhance your capacity for innovation. Technology-driven solutions allow you to manage production efficiently, reduce costs, and stay competitive:

- **Automated Growing Systems**: Automated systems that manage lighting, watering, and humidity can improve consistency and efficiency, allowing you to scale production while maintaining quality.

- **E-Commerce and Digital Sales**: Expanding into e-commerce enables you to reach customers outside your local area. A user-friendly online store and effective digital marketing allow you to broaden your reach and adapt to changing consumer shopping habits.

- **Data Analytics for Trend Insights**: Tracking customer preferences, sales patterns, and product performance can help you identify trends within your business. This information can guide future decisions and ensure you're meeting customer demands.

Conclusion

Successfully navigating the challenges and competition in the microgreens business requires a proactive, adaptable approach that addresses supply chain reliability, competitive pressures, and the constant need for innovation. By building resilient supply chains, diversifying your suppliers, and managing inventory with care, you can better withstand disruptions and ensure consistent production and delivery to your customers. These efforts help create a stable foundation for your business, allowing it to flourish even in uncertain conditions.

Understanding your competition and differentiating your brand is crucial to standing out in a crowded market. By closely analyzing competitors and identifying opportunities for unique offerings, you position your business as a distinctive choice, attracting customers who appreciate the quality and character of your microgreens. Brand loyalty often grows from consistent quality, trust, and a personal connection, so focusing on these areas gives your business a clear edge in a saturated market.

Staying innovative and responsive to trends is essential for long-term growth and relevance. By keeping a pulse on industry developments, experimenting with new products, and implementing efficient technologies, you keep your business agile and adaptable to changing customer needs. This commitment to innovation doesn't just help your business overcome challenges; it positions your brand as a leader in the microgreens market—a company that evolves with its customers and remains attuned to emerging opportunities.

With thoughtful planning, a commitment to quality, and a willingness to innovate, your microgreens business can continue to expand its reach, build lasting customer relationships, and maintain its place as a trusted, forward-thinking brand in the industry. Through resilience, strategic differentiation, and innovation, you create a solid path for long-term success in a competitive and ever-evolving market.

Chapter 12
Maintaining Business Growth

As your microgreens business expands, establishing efficient systems, building a skilled team, and exploring new revenue streams become essential components of sustaining growth. Scaling up production, managing increasing sales, and delivering top-tier customer service all require thoughtful planning and structured processes. Growth opens doors to exciting opportunities—broader market reach, higher revenues, and greater brand recognition—. Still, it also brings challenges in maintaining quality, consistency, and customer satisfaction as operations become more complex.

To grow effectively, it's important to lay a foundation to support increased demand without compromising the standards that set your brand apart. From streamlining production workflows to ensuring reliable inventory management, creating scalable systems allows your business to thrive under the pressure of expansion. Hiring and managing a capable team empowers you to delegate tasks confidently, maintain productivity, and nurture a positive work culture. Finally, diversifying revenue sources—whether through consulting, workshops, or online classes—enables you to reach new audiences and enhance financial stability.

In this chapter, we'll explore how to create efficient, scalable systems that support growing production and sales, build and manage a motivated team, and diversify your revenue streams to ensure steady, sustainable growth for your business. With these strategies, you'll be prepared to meet the demands of expansion while continuing to deliver exceptional quality and service to your customers.

Creating Systems to Manage Larger Production and Sales

Scaling your business requires systems that streamline operations and support higher production and sales volumes. As your customer base grows and orders increase, it's vital to have structured processes that maintain quality and efficiency. Developing systems helps you handle day-to-day operations and provides a foundation for further growth.

Standardizing Production Processes

Consistency is key to maintaining customer satisfaction, especially as demand grows. Standardized production processes ensure that every batch meets the same high standards and help your team follow reliable workflows as production scales.

- **Documenting Procedures**: Create detailed documents outlining each step of your growing process, from seed preparation to harvesting and packaging. Include guidelines on watering schedules, temperature and humidity control, and pest management. You can train employees consistently and maintain uniform quality across all production batches by having written procedures.

- **Implementing Quality Control**: Establish quality control checkpoints at key stages, such as germination, growth, and harvesting. Use these checks to catch issues early and ensure that only the highest-quality microgreens reach customers. Regular quality assessments help uphold your brand's reputation and reduce waste from crops that don't meet standards.

- **Utilizing Technology for Efficiency**: Consider investing in automated or semi-automated systems for lighting, irrigation, and climate control. Automation can reduce labor hours and improve consistency, especially when managing a larger volume of crops. Many tech solutions also allow remote monitoring, helping you keep tabs on growing conditions even when you're offsite.

Streamlining Sales and Inventory Management

Efficient sales and inventory management ensure customer orders are fulfilled accurately and on time, minimizing stockouts and oversupply. Managing sales becomes more complex as your customer base grows, so creating streamlined systems helps you maintain an organized, responsive business.

- **Inventory Tracking**: Implement a system to track inventory levels in real-time, allowing you to know exactly what's available for sale. This helps prevent overcommitting on orders and ensures that fresh microgreens are always available to meet demand. Simple software solutions or spreadsheets can be effective for small-scale tracking, while more advanced software can handle larger volumes.

- **Order Processing Systems**: An order management system efficiently tracks and processes customer orders. Many e-commerce platforms and point-of-sale (POS) systems offer integrated order management features, streamlining the process from order placement to fulfillment and invoicing. This approach reduces the likelihood of human error, ensuring orders are accurately filled and dispatched promptly.

- **Customer Relationship Management (CRM)**: As your customer base grows, it's helpful to have a CRM system to manage customer information, order history, and preferences. CRM software helps you provide personalized service, handle inquiries efficiently, and foster long-term relationships by keeping track of interactions and feedback.

Scaling Packaging and Delivery Logistics

Expanding production also means handling larger volumes of packaging and delivery, which can become complex as you serve more customers or venture into new markets.

- **Bulk Packaging Solutions**: Look for suppliers who can provide packaging in bulk to accommodate higher order volumes at a lower cost. Consider investing in eco-friendly, durable packaging options that reflect your brand values and accommodate larger quantities.

- **Outsourcing Delivery**: As deliveries increase, managing them all in-house may become difficult. Partnering with a local courier service or third-party logistics (3PL) providers can improve delivery efficiency and help you scale without overloading your team. Many 3PL providers offer same-day or next-day delivery services, which can keep your microgreens fresh for customers.

Hiring Employees and Managing Labor

As your business grows, hiring employees becomes essential to maintaining production, fulfilling orders, and managing operations. Building a capable, motivated team can be one of the most rewarding aspects of scaling a business, allowing you to delegate responsibilities and focus on strategic growth.

Determining Labor Needs

Understanding the tasks that require additional help helps you make strategic hiring decisions, ensuring you bring on employees who contribute to productivity and growth.

- **Assessing Roles and Responsibilities**: Start by identifying areas that need extra support, such as planting, harvesting, packaging, sales, or customer service. Determining which roles will benefit most from additional help can clarify your hiring priorities for full-time or seasonal roles during peak demand.

- **Part-Time vs. Full-Time Staff**: When deciding between part-time and full-time roles, consider your production volume and operational needs. For example, hiring part-time workers may be sufficient for labor-intensive tasks like harvesting. At the same time, a full-time operations manager could oversee quality control, supply orders, and other essential functions.

Recruiting and Training

Recruiting the right employees and providing thorough training helps your team consistently meet your standards, work efficiently, and contribute positively to your business.

- **Finding Qualified Candidates**: Advertise job openings through local job boards, community centers, or social media. Partnering with agricultural training programs or horticulture schools can help you connect with candidates with relevant skills.

- **Creating a Training Program**: Establish a structured training program that familiarizes new employees with your production methods, quality standards, and safety protocols. For example, a training guide with step-by-step instructions on planting, harvesting, and packaging ensures new hires understand and can follow your processes consistently.

- **Ongoing Skill Development**: Provide opportunities for skill development to keep your team engaged and knowledgeable. Cross-train employees in different areas of production, sales, or marketing to make your workforce more flexible and resilient. This approach improves productivity and helps employees feel valued and invested in the business.

Managing Employee Retention and Satisfaction

Keeping employees motivated and engaged is vital for retaining talent, minimizing turnover, and creating a positive work environment.

- **Offering Competitive Compensation**: Research wages for similar roles in your area and offer competitive pay to attract and retain skilled employees. Additionally, incentives like bonuses or profit-sharing based on team performance, which can increase motivation and loyalty, should be considered.

- **Creating a Positive Work Culture**: A supportive, inclusive work culture helps maintain high morale and reduces turnover. Foster open communication, recognize employees' contributions, and encourage a healthy work-life balance to keep your team satisfied and engaged.

- **Regular Feedback and Performance Reviews**: Schedule regular feedback sessions and performance reviews to discuss employees' progress, address challenges, and set goals. Constructive feedback helps employees improve while feeling valued and supported, contributing to a more productive and cohesive team.

Diversifying Revenue Streams: Consulting, Workshops, and Classes

To maintain sustainable growth and resilience, consider diversifying your revenue streams. Beyond selling fresh microgreens, you can generate additional income by offering related services, such as consulting, workshops, or classes. These activities allow you to share your expertise, reach new audiences, and further establish your brand as a trusted authority in the industry.

Consulting Services

Offering consulting services to other businesses or aspiring growers can be profitable, especially as demand for microgreens and sustainable farming practices increases.

- **Targeting Clients**: Potential clients include restaurants looking to grow microgreens on-site, schools interested in setting up educational gardens, or new growers seeking advice on production methods. Tailor your consulting services to each client's unique goals and needs.

- **Structuring Consulting Packages**: Create packages for different service levels, from basic setup consultations to ongoing support. For example, offer a one-time "Getting Started" package or a comprehensive "Monthly Advisory" service for ongoing guidance.

- **Building Authority**: By positioning yourself as an expert, you add credibility to your brand and attract clients who value your knowledge. Publish articles, speak at industry events, or share educational content online to showcase your expertise and grow your reputation in the industry.

Hosting Workshops and Classes

Workshops and classes generate revenue, help build community connections, and strengthen brand awareness. They offer customers a hands-on, educational experience that can deepen their interest in microgreens.

- **Identifying Popular Topics**: Focus on subjects that appeal to your target audience, such as "Growing Microgreens at Home," "The Health Benefits of Microgreens," or "Sustainable Gardening Practices." Understanding your audience's interests allows you to create workshops that attract attendees and generate engagement.

- **Offering Virtual and In-Person Options**: In-person workshops are excellent for building local connections, while virtual classes allow you to reach a broader audience. Video conferencing tools for online classes enable you to teach customers from anywhere while expanding your market.

- **Partnering with Local Venues**: Host workshops in collaboration with community centers, schools, and businesses. These venues can help you attract more attendees, increase visibility, and build relationships within the community.

Creating Educational Content and Online Courses

Online courses and digital content provide passive income opportunities and allow you to share your expertise at scale. Digital offerings make your knowledge accessible to people interested in microgreens and sustainable agriculture, even if they're outside your local area.

- **Developing Comprehensive Courses**: Design online courses that cover in-depth topics, such as "Advanced Microgreens Growing Techniques" or "Building a Sustainable Microgreens Business." Comprehensive courses can attract dedicated learners and provide a steady revenue stream.

- **Publishing E-books or Guides**: Consider creating e-books or downloadable guides on topics like microgreen varieties, nutrition, or business strategies. These resources can be sold on your website, generating income from customers who want in-depth information.

- **Utilizing Subscription Models**: Set up a subscription-based model to access premium content, such as exclusive videos, monthly Q&A sessions, or seasonal growing tips. Subscription models provide consistent revenue and build a loyal community of followers who value your ongoing expertise.

Conclusion

Sustaining business growth requires a thoughtful, structured approach that blends efficient systems, a motivated team, and diversified revenue streams. By implementing scalable processes across production, sales, and inventory management, you build a stable foundation to support increased demand without compromising the quality or consistency defining your brand. The strength of your team is equally vital; hiring skilled employees and cultivating a positive, collaborative work culture ensures your labor needs are met while keeping your team engaged, productive, and aligned with your vision.

Diversifying revenue sources through consulting, workshops, and online courses broadens your income and extends your reach. This allows you to connect with new audiences and reinforce your position as an authority in the microgreens industry. Each revenue stream complements your core business, providing stability and resilience even as market dynamics shift.

As your microgreens business grows, these strategies will equip you to meet the challenges of expansion while upholding the quality and customer care standards that have earned you a loyal following. Through strategic planning, thoughtful hiring, and creative revenue diversification, your business can scale sustainably, foster strong customer relationships, and stand out in an increasingly competitive market. By building on this solid foundation, your microgreens business can thrive today and adapt, innovate, and grow well into the future.

Chapter 13
Sustainable And Ethical Business Practices

As environmental awareness continues to rise, more consumers are choosing to support businesses that prioritize sustainable and ethical practices. Running a green business today goes beyond simply reducing environmental impact—it's about creating a positive, lasting influence on the community, embodying ethical responsibility, and fostering trust with customers. Embracing sustainability and ethical sourcing strengthens customer loyalty, enhances your brand's reputation, and positions your business as a force for good within the local community.

In this chapter, we'll explore practical ways to incorporate sustainable farming practices, select eco-friendly packaging, and build a community-driven business that supports local initiatives and practices ethical sourcing. By embedding these values into your microgreens business, you'll create a brand that resonates with conscious consumers and contributes to a more sustainable future.

Running a Green Business: Sustainable Farming Practices and Eco-Friendly Packaging

Operating a green business in the microgreens industry starts with implementing sustainable farming practices and selecting environmentally responsible packaging. These steps reduce your carbon footprint, minimize waste, and demonstrate a commitment to environmental stewardship that resonates with today's conscious consumers.

Sustainable Farming Practices

Sustainable farming practices are at the heart of a green microgreens business. As a grower, you can adopt methods that prioritize resource efficiency, soil health, and waste reduction, all of which contribute to long-term environmental benefits and appeal to eco-conscious customers. Here's how to make your farming practices more sustainable:

- **Minimizing Water Usage**: Microgreens can be water-intensive, but efficient watering systems, such as drip irrigation or wicking systems, can significantly reduce water waste. Hydroponic setups can be especially efficient for growing microgreens, using up to 90% less water than traditional soil-based methods by recirculating water and minimizing runoff.

- **Reducing Energy Consumption**: Indoor and greenhouse growers often rely on artificial lighting, but energy consumption can be minimized by using LED grow lights, which are more efficient and have a longer lifespan than traditional lights. In addition, choosing renewable energy sources, such as solar panels, further reduces reliance on fossil fuels and lowers your overall carbon footprint.

- **Composting and Waste Reduction**: Implementing a composting system for leftover plant material or organic waste from your farm reduces landfill waste and provides nutrient-rich compost that can be used to grow future crops. Composting turns waste into a valuable resource, completing a cycle that promotes soil health and reduces waste.

- **Using Organic and Natural Pest Control**: Rather than relying on chemical pesticides, which can harm the environment, focus on organic pest management strategies. Integrated Pest Management (IPM) is a sustainable approach that combines methods like crop rotation, biological pest control, and natural deterrents to manage pests with minimal environmental impact.

Eco-Friendly Packaging

Packaging plays a crucial role in how customers perceive your business's commitment to sustainability. Eco-friendly packaging reduces waste and aligns your brand with values that appeal to environmentally conscious customers.

- **Compostable and Biodegradable Packaging**: Compostable containers made from materials like cornstarch, paper, or bamboo allow customers to dispose of packaging without contributing to landfill waste. Clearly label packaging as compostable or biodegradable to raise awareness and encourage proper disposal.

- **Recyclable Materials**: If compostable options are not feasible, select packaging made from recyclable materials, such as PET or HDPE plastics. Use clear labeling to guide customers on how to recycle the packaging, and consider partnering with local recycling programs to support these efforts further.

- **Minimalistic and Efficient Design**: Reducing packaging to only what's necessary conserves resources and minimizes waste. Minimalistic packaging designs that avoid excess plastic, inks, or non-recyclable materials reduce environmental impact and convey a clean, modern aesthetic that appeals to many customers.

- **Reusable Packaging Options**: For customers who purchase microgreens regularly, consider offering a reusable packaging option, such as jars or containers they can return for refills. This approach fosters customer loyalty and allows you to reduce single-use packaging waste significantly.

By adopting sustainable farming practices and eco-friendly packaging, you position your business as an environmentally responsible brand. These efforts contribute to a healthier planet and attract customers who prioritize supporting green businesses.

Building a Community-Driven Business: Supporting Local Initiatives and Ethical Sourcing

Running a sustainable and ethical business is about more than environmental responsibility—it's also about building connections within your community and prioritizing fair, ethical sourcing. A community-driven business cultivates meaningful relationships with customers and neighbors, supports local initiatives, and sources products in ways that reflect high ethical standards.

Supporting Local Initiatives

Supporting local initiatives allows your business to contribute positively to your community, fostering goodwill and building strong relationships with local customers. Here are ways to make a positive impact:

- **Partnering with Local Organizations**: Collaborate with local non-profits, schools, or food banks to support projects that align with your values. For example, donating microgreens to a food bank or offering nutrition education workshops at schools supports the community and introduces new people to your product.

- **Hosting Community Events and Workshops**: Hosting workshops on sustainable farming, composting, or healthy eating encourages community members to engage with your brand and learn valuable skills. These events create a positive association with your business and establish you as a helpful, involved community member.

- **Giving Back Through Local Programs**: Consider setting up a donation-based program where a portion of each sale supports a community cause. Giving back regularly demonstrates a long-term commitment to your community's welfare, making customers feel good about supporting your business.

- **Creating a Loyalty Program for Local Customers**: A loyalty program that offers discounts or rewards to local customers encourages repeat business and fosters a sense of community. Rewarding local customers with perks shows appreciation for their support and keeps your business closely tied to the community.

Ethical Sourcing Practices

Ethical sourcing is vital to running a community-focused and socially responsible business. Sourcing your seeds, growing media, and other materials responsibly reflects your commitment to fair labor, environmental responsibility, and local support.

- **Sourcing Locally When Possible**: Purchasing seeds, supplies, and packaging from local or regional suppliers reduces your carbon footprint by minimizing transportation and supporting other local businesses. Local sourcing also allows for stronger supplier relationships and greater transparency about production methods.

- **Partnering with Fair Trade and Ethical Suppliers**: When local options aren't available, prioritize suppliers that follow fair trade practices or ethical standards, especially for international materials. Fairtrade-certified suppliers ensure that workers receive fair wages, work in safe conditions, and adhere to environmentally responsible practices.

- **Transparency with Customers**: Today's consumers value transparency, so be open about where and how your products are sourced. For example, include information about your suppliers on your website or packaging, highlighting your commitment to ethical practices. Transparency fosters trust and helps customers make informed choices about supporting your business.

- **Promoting Diversity and Inclusivity in Sourcing**: When choosing suppliers, consider supporting suppliers from diverse backgrounds, such as women-owned or minority-owned businesses. Promoting inclusivity within your supply chain reinforces your commitment to fair practices and creates a more resilient, supportive network.

By actively supporting local initiatives and practicing ethical sourcing, you create a business prioritizing community welfare and social responsibility. These efforts strengthen your brand, foster loyal customer relationships, and create a positive social impact that resonates with conscious consumers.

Embracing a Holistic Approach to Sustainable and Ethical Practices

Sustainability and ethical practices extend beyond isolated efforts—they involve a holistic approach that touches every aspect of your business. By embedding these values into your mission, operations, and interactions with the community, you create a brand that is not only environmentally and socially responsible but also genuinely inspiring to your customers.

Creating a Sustainability Mission Statement

A sustainability mission statement communicates your commitment to green and ethical practices to customers, employees, and partners. This statement acts as a guiding principle for decision-making and a public commitment to your values:

- **Define Your Core Values**: Identify the values that matter most to your business, whether it's environmental stewardship, supporting local communities, or ethical sourcing. Clearly define these priorities in your mission statement.

- **Set Achievable Sustainability Goals**: Setting specific goals gives you and your team concrete objectives to work toward. For example, aim to reduce water usage by 20% or transition entirely to biodegradable packaging within a year. Goals demonstrate that your commitment to sustainability is actionable and measurable.

- **Communicate Transparently with Customers**: Share your mission statement and sustainability goals with customers through your website, packaging, and social media channels. Providing updates on progress and celebrating milestones can help keep customers engaged and invested in your journey.

Training Employees in Sustainable and Ethical Practices

Engaging employees in your sustainability mission ensures that green and ethical practices are consistently applied across the business, from production to customer interactions.

- **Integrate Training into Onboarding**: Include sustainability training in onboarding so every new team member understands your practices and values. Provide practical training on topics like energy conservation, composting, or minimizing waste in daily operations.

- **Encourage a Culture of Responsibility**: Foster a work environment where employees feel empowered to suggest improvements or take ownership of sustainable practices. Encourage team discussions about sustainability ideas and recognize employees who go above and beyond to support green initiatives.

- **Provide Resources for Learning**: Offer resources, such as workshops, online courses, or informative materials, for employees to learn more about sustainability and ethical business practices. Continuous education helps employees stay informed about industry advancements and new sustainable practices.

Conclusion

Building a microgreens business grounded in sustainable and ethical practices is more than responding to market trends—it's about shaping a company that makes a positive, lasting impact on the environment, strengthens the local community, and embodies values of responsibility and integrity. By incorporating sustainable farming practices, selecting eco-friendly packaging, and actively supporting community initiatives, your business reduces its environmental footprint while fostering meaningful connections within your community. Ethical sourcing reinforces this commitment, ensuring your supply chain aligns with fair labor, environmental care, and inclusivity principles.

A holistic approach to sustainability—supported by a clear mission statement and a dedicated team—positions your microgreens business as a leader in responsible entrepreneurship. As your business grows, these core values will serve as a guiding compass, empowering you to make decisions that reflect care for the planet and the people you serve. This approach builds trust and inspires customers who are increasingly seeking brands that prioritize both environmental and social responsibility.

Through thoughtful, community-focused, and ethically conscious practices, your microgreens business will flourish as a trusted, influential presence in the industry. By holding sustainability and ethics at the heart of your business, you create a brand that meets the needs of today's consumers and contributes to a better, more sustainable future for all.

Chapter 14
Microgreens: Varieties, Growing and Harvesting

This section is a guide to many varieties of microgreens, their planting, growing, and harvesting needs, and a brief flavor profile of over thirty popular microgreens.

Many of the microgreens featured here grow in less than two weeks and are best harvested at the cotyledon stage just as the first tiny true leaves are beginning to emerge. Letting them develop past this time will cause them to develop larger, more mature true leaves, which are exciting but can change the greens' flavor or texture. For instance, sunflower shoots can taste quite bitter at this stage, while they are sweet and nutty at the cotyledon stage. This guide gives approximate development time frames for reaching the cotyledon stage; however, growing conditions can change the timeline. Please experiment and tweak based on your preferences and growing environment.

Key Definitions

Ease of Growth refers to the beginner-friendliness of various seeds. It is important to recognize that some microgreens are more forgiving and easier to grow than others, especially for those just starting their microgreen adventures.

Seeding Density explains the optimal seed density to sow per tray. This ensures a thick, lush harvest while ensuring the seeds have ample space to flourish without overcrowding. A kitchen scale is the best tool for this.

PreSoak: some seeds require soaking to soften the seed coat before sowing. Some seeds require no presoak, while others require extensive soaking. This section will give detailed instructions about how best to soak the seeds.

Germination Rate refers to the percentage of seeds expected to germinate under ideal conditions. It is a crucial metric for growers because it indicates the viability and quality of seeds. A high germination rate suggests that many seeds are healthy and capable of growing under appropriate conditions, making them more reliable for planting.

Germination Time: Germination time refers to the period it takes for a seed to break dormancy and start its growth process, culminating in the emergence of a sprout from the soil or growing medium.

Harvest Time: Treat this number as a guideline. Observing the plants for the appearance of true leaves, rather than strictly adhering to a predetermined number of days, is a good practice for determining the best time to harvest.

Microgreens Flavor: Understanding the flavor profiles of different microgreens allows cooks to select the suitable types to complement and elevate their culinary creations, whether looking for a subtle hint of freshness or a bold flavor accent.

Varieties of Microgreens

Amaranth: This small, delicate microgreen prefers warmer growing conditions; however, it can be sensitive to too much light. This delicate green does not grow very tall, so the resulting leaves will need to be harvested close to the soil line and rinsed well before consumption.

Ease of Growth: moderately difficult: grow in warmer conditions
Seeding Density: 15 grams or 0.5 oz. per 10 x 20 x 1" tray
Presoak: no
Germination Rate: high

Germination Time: 2 to 3 days

Harvest time: 12 to 14 days

Color: ranges from light green to pink or dark red

Microgreens Flavor: mild, earthy

Arugula: This one has a slower growth rate and is harder to grow than other micros; however, the peppery pop it gives is worth the extra effort. The peppery taste gets milder as the greens mature, so harvest accordingly. Arugula does not like direct light.

Ease of Growth: moderately difficult

Seeding Density: 12 grams or 0.4 oz. per 10 x 20 x 1" tray

Presoak: no

Germination Rate: high

Germination Time: 3 days

Harvest time: 6 to 12 days

Color: green

Microgreens Flavor: peppery

Basil: This includes Dark Purple, Genovese, and Thai basil. Keep basil seeds damp with regular misting until sprouted and roots are established. Basil is easy to grow, but you may find it grows more slowly than other microgreens.

Ease of Growth: easy

Seeding Density: 10 grams or 0.35 oz per 10x20 tray

Presoak: no

Germination Rate: high

Germination Time: 3 to 4 days

Harvest time: 20 to 25 days (earlier harvest may be possible in good growing conditions.)

Color: Thai and Genovese: green; Dark Opal: purple with some green

Microgreens Flavor: Purple and Genovese: intense basil; Thai basil with a distinct hint of anise

Beets: Harvest close to the soil to keep the colorful red stem intact. Makes a beautiful garnish. Sow soaked seeds thickly across compacted soil; cover with a thin layer of soil, then tamp down lightly for best results.
Ease of Growth: very easy
Seeding Density: 20-30 grams or 1 oz per 10 x 20 x 1" tray
Presoak: 4 to 8 hours in cold water
Germination Rate: good
Germination Time: 3 to 4 days
Microgreens Harvest time: 10 to 14 days
Microgreens Color: green leaves with dark red stems
Microgreens Flavor: earthy

Broccoli is perhaps the fastest and easiest microgreen to grow, so it's great for beginners. It can be sown a little more densely than other greens. This hardy and substantial microgreen makes a great base for a microgreen salad.
Ease of Growth: very easy
Seeding Rate: 15-20 grams or 0.5 oz per 10 x 20 x 1" tray
Presoak: no
Germination Rate: high
Germination Time: 2 to 3 days
Harvest time: 8 to 12 days
Microgreens Color: green
Microgreens Flavor: mild cabbage

Brussels Sprout: The Long Island Brussels Sprout variety grows very fast and has a mild flavor. Brussels sprouts require a slightly cooler temperature for germination.

Ease of Growth: easy

Seeding Density: 15-20 grams or 0.5 oz per 10 x 20 x 1" tray

Presoak: no

Germination Rate: high

Germination Time: 1 to 2 days

Harvest Time: 8 to 10 days

Color: dark green

Microgreens Flavor: mild Brussels sprout

Cabbage: All cabbage varieties are easy to grow, very tasty, and make a great base for a microgreens salad. The different varieties offer great color combinations. More detailed information on cabbage follows the growing information.

Ease of Growth: easy

Seeding Rate: 15-20 grams or 0.5 - 0.7 oz. Per 10 x 20 x 1" tray.

Presoak: no

Germination Rate: high

Germination Time: 1 to 2 days

Harvest time: 8 to 12 days

Color: see below

Microgreens Flavor: mild cabbage

Varietal differences:

Golden Acre: green with slightly yellow tint

Red Acre & Red Rock Mammoth: Green leaves with purple highlights and stems.

Red Rock Mammoth typically has larger leaves.

Bok Choi (Pak Choi, Pok Choy): These are pale or dark green with white stems. They are more sensitive to light, so they should be kept in low-light conditions after the third or fourth day.

Purple Bok Choy: purple-topped leaves with green underside.

Kogane or Pekka Santos: Pale yellowish green. Mild choy flavor.

Cauliflower is easy to grow and can be sown more thickly. It makes a great base for a microgreens salad.

Ease of Growth: easy
Seeding Density: 15-20 grams or 0.5 oz per 10 x 20 x 1" tray
Presoak: no
Germination Rate: high
Germination Time: 1 to 2 days
Harvest time: 8 to 10 days
Color: green leaves and violet stems
Microgreens Flavor: mild

Celery: This very pale green microgreen is lovely with chicken salad or added to scrambled eggs. It has a long germination time. Celery needs lots of light to grow well, and a grow light works better than sunlight for this.

Ease of Growth: moderately easy
Seeding Density: 15-20 grams or 0.5 oz per 10 x 20 x 1" tray
Presoak: no
Germination Rate: high but can take time
Germination Time: up to 2 weeks
Harvest time: 2 to 5 weeks post-germination
Color: light green
Microgreens Flavor: intense celery

Chard: Swiss or Rainbow varieties work equally well as microgreens. Amazing color assortment in rainbow chard.

Ease of Growth: easy
Seeding Density: 20-30 grams or 1 oz per 10 x 20 x 1" tray

Presoak: 4 to 8 hours in cold water
Germination Rate: good
Germination Time: 3 to 4 days
Harvest time: 10 to 14 days
Color: Rainbow: green leaves, multi-colored stems. Ruby Red: green leaves and red stems
Microgreens Flavor: chard/spinach

Chive/Green Onion/Leek: These are slow growers but can be kept growing for a second or even third harvest. These latter harvests tend to be more flavorful than the first. Some growers find that leeks grow faster than onions or chives.
Ease of Growth: easy
Seeding Density: 40-50 grams or 1.5 oz. per 10 x 20 x 1" tray
Presoak: no
Germination Rate: high
Germination Time: 1 to 2 weeks
Harvest time: 3 weeks
Color: green
Microgreens Flavor: onion

Cilantro: Also called Chinese Parsley. Use split seeds because they germinate faster. Germination time can vary, with some seeds germinating up to a week after the first. Sow thickly; cover with about a 1/4" layer of soil; tamp gently.
Ease of Growth: relatively easy
Seeding Density: 30-40 grams a little over 1 oz. per 10 x 20 x 1" tray
Presoak: 2-4 hours
Germination Rate: high
Germination Time: 1 to 2 weeks
Harvest time: 3 to 4 weeks

Color: green

Microgreens Flavor: intense cilantro/coriander

Collard: The easiest microgreen variety to grow is Vates

Ease of Growth: varies with variety, but Vates is an easy grower

Seeding Density: 15-20 grams or 0.5 oz per 10 x 20 x 1" tray

Presoak: no

Germination Rate: high

Germination Time: 2 to 3 days

Harvest time: 6 to 10 days

Color: dark green

Microgreens Flavor: fresh, intense collard green flavor

Curly Cress: Needs significantly less water in the sprouting stage than other seeds. Too much water equals a low germination rate. Mist up to three times a day, with less mist each time. Needs to be damp, but not soggy.

Ease of Growth: fairly easy

Seeding Density: 20 grams or 0.7 oz per 10 x 20 x 1" tray

Presoak: No (mucilaginous)

Germination Rate: high

Germination Time: 3 to 5 days

Harvest time: 8 to 12 days

Color: green

Microgreens Flavor: intense peppery, with a hint of sweet.

Fennel: Sow a little more thinly to allow room for seeds to grow. Seed husks tend to cling to the leaves, so plan on growing long enough for husks to drop.

Ease of Growth: fairly easy

Seeding Density: 20-30 grams or 1 oz. per 10 x 20 x 1" tray

Presoak: no

Germination Rate: high
Germination Time: 3 to 4 days
Harvest time: 16-20 days
Color: green
Microgreens Flavor: mild, anise flavor

Kale: Includes both Curly and Red Russian.
Ease of Growth: easy
Seeding Density: 15-20 grams or 0.5 oz per 10 x 20 x 1" tray
Presoak: no
Germination Rate: high
Germination Time: 2 to 3 days
Harvest time: 6 to 10 days
Color: see below
Microgreens Flavor: fresh kale

Blue Curled Kale microgreens are tinged with blue and have a slight curl to the leaf; Red Russian Kale has a blush of red on the underside of the leaf.

Mint: Also called Micro Mint
Eat of Growth: easy
Seeding Density: 15 to 20 grams per 10 x 20 x 1" tray
Presoak: no
Germination Rate: high
Germination Time: 3 to 4 days
Harvest Time: 7-14 days
Color: g reen

Microgreens Flavor: crisp mint flavor, slightly herbaceous and sweet.

Mustard: Includes Mizuna, Red Streaks, Osaka, Southern Giant, Crimson Tide, Mibung and Yellow

Ease of Growth: fairly easy
Seeding Density: 10-15 grams or 0.35-0.5 oz per 10 x 20 x 1" tray
Presoak: No
Germination Rate: high
Germination Time: 2 to 3 days
Harvest time: 8 to 12 days
Color: green to red, depending on type
Microgreens Flavor: spicy mustard

Parsley: Seeds take a bit longer to germinate than other microgreens.
Ease of Growth: easy
Seeding Density: 15-20 grams or 0.5 oz per 10 x 20 x 1" tray
Presoak: No
Germination Rate: high
Germination Time: 6-7 days
Harvest time: 16+ days
Color: green
Microgreens Flavor: mild parsley

Pea Shoots: Dun or Afila Tendril. Use plenty of water when soaking; Spread evenly but thickly when sowing; tamp the soil gently. Mist once a day. Keep soil damp but not soggy. One of the more challenging microgreens to grow.
   Ease of Growth: difficult
   Seed Density: 200-275 grams or 7-9 oz. per 10 x 20 x 1" tray
   Presoak: 6-12 hours (cold water)
   Germination Rate: high
   Germination Time: 2 to 3 days
   Harvest time: 8 to 12 days
   Color: green

Flavor: crunchy, mildly sweet, fresh

Radish: Daikon or Sango. It grows fast and can be ready to harvest as early as 5 days. Reach peak crunchiness about day 5 or 6. First, leaves can be spiky, so harvesting in the cotyledon stage is recommended.

Ease of Growth: easy
Seeding Density: 30-35 grams or 1 oz. per 10 x 20 x 1"tray
Presoak: no
Germination Rate: high
Germination Time: 1 to 2 days
Harvest time: 5 to 12 days
Color: Sango is a beautiful shade of purply green; Daikon is green
Microgreens Flavor: strong radish

Sunflower: Black Oil. Pre-sprouting yields better results. See Note for instructions.

Ease of Growth: requires more seed prep than other microgreens. Moderately difficult to grow, but worth the effort.

Seeding Density: 250 grams or about 9 oz. per 10 x 20 x 1" tray
Presoak: 6-12 hours pre-sprout 12-24 hours in colander (see below)
Germination Rate: high
Germination Time: 2 to 3 days
Harvest time: 8 to 12 days
Color: green
Microgreens Flavor: crunchy, nutty, fresh

Note: Soak seeds for 6-12 hours; then transfer to a wire mesh strainer for sprouting. Rinse 2 to 4 times a day for a day or two or until you can see the seedling beginning to sprout. Transfer to the growing tray; tamp gently to ensure the seeds are in good contact with the soil. Mist twice daily; do not expose to light until 3 days after sprouting. Keep soil moist but not soggy.

Wheatgrass is prone to mold, so keep the soil dry but moist. For best results, it needs cooler temperatures, under 75°F.

Ease of Growth: moderately easy

Seeding Density: 450 g. or about 16 oz of seed for a 10 x 20 x 1" tray

Presoak: (4-8 hrs) do not soak for more than 8 hrs.

Germination Rate: high

Germination Time: 2-3 days

Harvest time: 8-10 days

Color: green

Microgreens Flavor: fresh green

Chapter 15
Bonus: Business Tools And Resources

Growing a successful microgreens business requires effective management, planning, and continual learning. With the right tools and resources, you can streamline operations, gain valuable insights, and stay ahead of market trends. This bonus section provides:

- A curated list of helpful apps and software for managing inventory, accounting, and marketing.

- Recommended books, websites, and podcasts for business development.

- Practical templates to support your business planning, cost tracking, and more.

- Essential Apps and Software for Small Business Management

Using the right technology can simplify daily tasks, help you manage finances, and improve customer engagement. Here are some valuable tools for small business owners in the microgreens industry:

Inventory Management

- **Sortly**: This app offers visual inventory tracking and barcode scanning, making it ideal for managing growing supplies and packaging materials.

Its intuitive interface makes it easy to track quantities and locations.

- **Zoho Inventory**: A robust inventory management software for small businesses that includes order tracking, low-stock alerts, and integration with e-commerce platforms. Zoho also integrates well with Zoho Books for accounting.

- **Square for Retail**: Great for businesses with in-person and online sales, Square for Retail helps you manage inventory, track sales, and accept payments in one platform.

Accounting and Financial Management

- **QuickBooks Online**: QuickBooks is a versatile accounting solution that helps with everything from expense tracking and invoicing to payroll and tax preparation. It's widely used and offers various integrations with other business software.

- **Wave**: A free accounting tool for small businesses that includes invoicing, expense tracking, and receipt scanning. It's simple to use and ideal for small-scale businesses without complex accounting needs.

- **Expensify**: Useful for tracking receipts and managing expenses, Expensify can help you stay organized and reduce paperwork. It's particularly helpful for monitoring supply purchases and travel expenses.

Marketing and Social Media Management

- **Canva**: Canva is a design tool for creating marketing materials, social media posts, and even website graphics. With templates and easy-to-use features, it's perfect for building a consistent brand look.

- **Buffer**: A social media management platform that allows you to schedule and analyze posts across multiple social networks. Buffer simplifies managing marketing content and helps you maintain a steady online presence.

- **Mailchimp**: This platform is known for email marketing but also offers tools for creating landing pages, automating customer outreach, and tracking marketing performance. Mailchimp's automation features make it easy to stay connected with customers.

Project Management and Collaboration

- **Trello**: An intuitive project management tool, Trello uses a card-based system to help you organize tasks, collaborate with team members, and keep track of projects, such as planning harvest cycles, tracking deliveries, or organizing marketing campaigns.

- **Asana**: Asana is a versatile task management app that allows for detailed project planning, deadline tracking, and team collaboration. It's a great choice for managing daily operations and keeping everyone on the same page.

- **Slack**: Slack is a communication tool that keeps your team connected through organized channels. It's especially helpful for remote or hybrid teams, making it easy to discuss projects, share files, and stay up to date.

Recommended Books, Websites, and Podcasts for Business Development

Continual learning is key to growing a thriving business. These resources provide insights, inspiration, and practical advice on topics ranging from small business strategies to sustainable agriculture.

Books

- **"The Lean Startup" by Eric Ries**: This book covers startup methodologies, including tips for testing ideas quickly and adjusting based on customer feedback. It's valuable for new business owners looking to build a sustainable, customer-focused company.

- **"Profit First" by Mike Michalowicz**: This unique approach to financial management teaches business owners how to prioritize profit through practical cash flow management strategies.

- **"The Market Gardener" by Jean-Martin Fortier**: This book offers insights on running a profitable small-scale farm with sustainable practices. It's especially relevant for microgreens businesses focused on efficient, eco-friendly growing.

- **"Building a StoryBrand" by Donald Miller**: This book provides a framework for crafting a compelling brand narrative that resonates with customers. It's essential for learning how to communicate your business's value effectively.

Websites

- **Urban Ag News (www.urbanagnews.com):** Urban Ag News offers industry news, research, and trends focused on urban agriculture and controlled-environment farming. It is a great source for microgreens growers interested in industry developments.

- **LocalHarvest (www.localharvest.org)**: This website helps small farms connect with consumers looking for fresh, local produce. You can list your microgreens business here to increase visibility among local buyers.

- **Extension.org**: The U.S. Cooperative Extension website provides sustainable agriculture, food safety, and business management resources. It's an excellent source of free resources and advice for small-scale farmers.

Podcasts

- **"The Modern Acre"**: Focused on agriculture and entrepreneurship, "The Modern Acre" podcast covers topics like farm management, sustainability, and innovative business practices with guest experts and industry leaders.

- **"Small Farm Nation"**: Hosted by small farming expert Tim Young, this podcast shares tips on branding, marketing, and creating a profitable small farm. It's valuable for learning how to build a loyal customer base and scale operations.

- **"How I Built This" by NPR**: This popular podcast explores the journeys of entrepreneurs in various fields, sharing insights on resilience, innovation, and business growth. While not agriculture-specific, it's an inspiring listen for any business owner.

Templates for Business Planning, Cost Tracking, and More

Well-designed templates can simplify planning, tracking, and organizing key areas of your business. Below are suggested sections for creating effective templates tailored to your specific needs. These guidelines will help you design templates that keep you organized and support data-driven decisions. By building custom templates that reflect your unique business requirements, you'll gain tools that streamline daily operations and keep you focused on growth.

Business Planning Template

A business plan template helps clarify your goals, target market, and financial projections. This template may be structured to guide you through each section of a business plan, including:

- **Executive Summary**
- **Market Analysis**
- **Product Offerings**
- **Marketing Strategy**
- **Financial Projections**

A structured template makes presenting your business plan to potential investors, lenders, or partners easier.

Cost Tracking Template

A cost-tracking template lets you monitor expenses for growing, packaging, and selling your microgreens. This template typically includes:

- **Material Costs (seeds, soil, water)**
- **Labor Costs**
- **Packaging and Delivery Costs**
- **Monthly Fixed Costs (rent, utilities)**

- **Total Cost per Product Unit**

Tracking these costs regularly allows you to adjust pricing, identify areas for cost reduction, and better understand your profit margins.

Inventory Management Template

An inventory management template keeps your supplies and stock organized, allowing you to track the quantities and costs of seeds, growing materials, and packaging. A well-organized inventory sheet includes:

- **Item Name and Description**
- **Quantity in Stock**
- **Reorder Level**
- **Cost per Unit**
- **Supplier Contact Information**

Creating a template helps you avoid stockouts, manage supply chains efficiently, and streamline reordering.

Sales and Revenue Tracking Template

A sales and revenue tracking template lets you monitor your monthly, quarterly, and yearly sales performance. Key sections might include:

- **Date of Sale**
- **Product Sold**
- **Sales Channel (online, farmers market, wholesale)**
- **Revenue from Each Sale**

- **Monthly and Cumulative Revenue**

Tracking revenue helps you assess the performance of different products and sales channels, identify trends, and make informed decisions on scaling production.

Marketing Calendar Template

A marketing calendar template organizes your campaigns and promotional activities, helping you plan posts, events, and seasonal promotions. Sections often include:

- **Date**
- **Campaign/Event Description**
- **Marketing Platform (social media, email, etc.)**
- **Target Audience**
- **Budget**

Having a clear schedule enables you to create a consistent marketing rhythm, effectively engage your audience, and track results for improvement.

Final Thoughts

With the right tools, resources, and customized templates, you're now well-prepared to streamline operations, deepen your business knowledge, and set the foundation for lasting success. Running a microgreens business demands organization, adaptability, and a commitment to continual learning—qualities that these resources support by helping you stay efficient, informed, and focused on growth.

As you move forward, remember that building a sustainable and thriving business is an ongoing journey. Each new tool, book, and connection you discover equips you to handle challenges and inspires fresh ideas and innovations. Embrace each resource as an opportunity to refine your approach, expand your skills, and connect with others who share your values.

By dedicating time to hone these practices, you're creating a profitable business and cultivating a brand rooted in quality, responsibility, and resilience. Enriched by continuous learning and thoughtful planning, this journey paves the way for a microgreens business that can grow, evolve, and impact your community and beyond. The road ahead is full of potential, and with each new step, you're building a meaningful legacy in sustainable agriculture.

Chapter 16
Conclusion
A Path To A Thriving Microgreens Business

As you reflect on the journey of building a successful microgreens business, it's clear that this industry offers far more than just the opportunity to grow and sell plants. Running a microgreens business can be fulfilling and profitable, blending creativity, sustainability, and purpose. It's a venture beyond commerce, allowing you to contribute to a healthier, more sustainable food system. Whether you were drawn to this path by a passion for fresh, nutritious food, a desire to make a positive environmental impact, or the dream of entrepreneurship, this business offers a unique space to create something of real, lasting value.

Each phase of growth—selecting the best varieties, scaling production, mastering marketing, and embedding sustainable practices—presents new learning experiences, challenges, and rewards. You create a brand that resonates with today's conscientious consumers by cultivating a business that aligns with your values, serves customer needs, and builds meaningful community connections. This alignment fosters customer loyalty and positions your business for long-term success.

As the microgreens industry evolves, staying committed to growth, learning, and responsiveness will keep your business relevant, adaptable, and resilient. The demand for nutrient-dense, sustainable foods shows no signs of slowing, and by embracing innovation, you ensure that your brand stays ahead of market trends. With each step forward, you're shaping a thriving business and contributing to a more health-conscious, environmentally friendly future—an accomplishment that brings personal and professional fulfillment.

Final Thoughts on Innovation and Growth

One of the keys to a thriving microgreens business is a mindset of continuous innovation and improvement. As you move forward, embrace opportunities to refine your processes, explore new varieties, and diversify your offerings to stay ahead of industry trends. The market for fresh, nutrient-rich, and sustainably grown food is expanding rapidly, and those who adapt and innovate will be the ones who thrive. Experimenting with value-added products, finding unique ways to connect with your customers, or even exploring partnerships with other local businesses can keep your business dynamic and engaging. Every new idea and experiment—whether in packaging, marketing, or product offerings—can set you apart and further build your brand.

The journey will be challenging, but staying committed to your vision, adapting to challenges, and continuously learning will help your business flourish. Remember, each step to improve your processes, grow your skills, or expand your market reach brings you closer to creating a legacy of success, sustainability, and customer loyalty.

Resources for Continued Learning and Support

The microgreens industry, like all of agriculture, is constantly evolving. Staying informed, networking with other growers, and accessing the latest resources will help you stay ahead. Here are some resources to support your continued learning and business development:

- **Industry Publications and Journals**: *Urban Ag News*, *Greenhouse Grower*, and *AgFunderNews* regularly publish insights, research, and trends relevant to microgreens and small-scale agriculture. These can help you stay informed about industry advancements, market trends, and sustainable practices.

- **Agricultural Extension Programs**: Many local agricultural extension programs offer free or affordable resources, training, and networking opportunities for small-scale growers. Extension offices can provide expertise in pest management, crop selection, and sustainable practices.

- **Professional Organizations and Associations**: Joining organizations like the Specialty Food Association or the American Horticultural Society connects you with a network of industry professionals, conferences, and events. These memberships often include access to exclusive research, trade shows, and industry certifications that can strengthen your expertise.

- **Online Courses and Certifications**: Platforms like Udemy, Coursera, and FutureLearn offer affordable courses in horticulture, sustainable agriculture, and small business management. Investing in courses can help you refine specific skills, whether you want to master hydroponics, understand soil health, or develop advanced marketing techniques.

- **Social Media and Online Communities**: Connecting with other microgreens growers and small business owners through social media groups and online forums can be an invaluable source of support and inspiration. Platforms like Facebook, Reddit, and Instagram host thriving communities where members share tips, experiences, and troubleshooting advice. Joining these groups allows you to learn from others in the field, share your successes, and keep up with industry trends.

- **Mentorship Programs and Local Business Networks**: Many regions offer mentorship programs for small business owners, such as SCORE, which pairs entrepreneurs with experienced business mentors. Additionally, local business networks or chambers of commerce offer networking, collaboration, and business development opportunities.

By continually learning and surrounding yourself with resources, mentors, and peers, you will be better prepared to navigate the challenges and seize the opportunities that come your way.

A Last Word: Growing Your Vision

Building a microgreens business is more than just a career path—a journey deeply rooted in passion, purpose, and immense potential. Your dedication to high-quality products, sustainable practices, and community-centered values can create ripples beyond your business, inspiring others and contributing to a future that values local agriculture, environmental stewardship, and mindful, health-focused lifestyles. Every choice you make, from selecting eco-friendly practices to building lasting customer relationships, shapes your brand and the landscape of food production and conscious consumerism.

As you move forward, embrace each new opportunity to innovate, learn, and make a difference. This journey invites continuous growth and creativity, and remaining adaptable allows you to evolve with changing times and new market demands. Each step you take—introducing a new product, refining your processes, or impacting your community—reinforces your vision and strengthens your brand's foundation. Take pride in your business's positive contributions to your customers, community, and planet, as these will become an integral part of your legacy.

With a clear vision, a commitment to your values, and a willingness to keep exploring new ideas, there's virtually no limit to what your microgreens business can achieve. This journey is uniquely yours to shape, offering endless possibilities to grow, inspire, and succeed. The future is bright with opportunity, and the impact of your work has the potential to create lasting, meaningful change in the lives of your customers and your community. Let this inspire you to keep growing—a vibrant, thriving business awaits, driven by purpose and vision.

Chapter 17
References

Books

1. This book provides practical advice on financial management, essential for maintaining profitability in a small business.

2. **Fortier, J.-M.** (2014). *The Market Gardener: A Successful Grower's Handbook for Small-Scale Organic Farming*. New Society Publishers.

3.

4. Focuses on sustainable and profitable small-scale farming, offering insights for microgreens growers looking to optimize their methods.

5. **Miller, D.** (2017). *Building a StoryBrand: Clarify Your Message So Customers Will Listen*. HarperCollins Leadership.

6.

7. A guide to building a compelling brand narrative, helpful for business owners looking to strengthen customer connections through effective storytelling.

8. **Holmgren, D.** (2002). *Permaculture: Principles and Pathways Beyond Sustainability*. Holmgren Design Services.

9.

10. This book explores sustainability and permaculture principles, relevant for microgreens businesses committed to eco-friendly practices.

Websites and Online Resources

1. Provides a directory for small farms, connecting growers with consumers who value fresh, local produce.

2. **U.S. Cooperative Extension (Extension.org)** - www.extension.org

3.

4. Free resources, research, and expert advice on agriculture, sustainable practices, and business management are offered.

5. **Greenhouse Grower** - www.greenhousegrower.com

6.

7. Industry insights and practical resources for greenhouse growers, including those focused on small-scale, sustainable agriculture.

8. **Small Business Administration (SBA)** - www.sba.gov

9.

10. The SBA provides resources on business planning, financial management, and funding options tailored for small businesses.

Podcasts

1. **The Modern Acre** - Covers topics related to modern agriculture, entre-

preneurship, and sustainability with interviews from industry experts.

2. **Small Farm Nation** - Hosted by Tim Young, this podcast offers insights into marketing, branding, and profitable farm management.

3. **How I Built This by NPR** - A podcast that explores entrepreneurial journeys, sharing lessons on resilience, innovation, and business growth.

Articles and Journals

1. **Journal of Sustainable Agriculture—This journal focuses on** agricultural practices on, research and practices in sustainable agriculture, with articles on crop production, pest management, and soil health.

2.

3. Example: *"Sustainable Practices for Small-Scale Organic Farms,"* Journal of Sustainable Agriculture, 2019.

4. **Permaculture Research Institute** - Articles and resources on permaculture principles, soil health, and sustainable agriculture practices.

5.

6. Example: *"Building Healthy Soil in Small-Scale Farms,"* Permaculture Research Institute, 2021.

Recommended Templates and Tools

1. **QuickBooks** - Comprehensive accounting valuable software for tracking expenses, invoicing, and financial planning.

2. **Canva** - Design tool for creating branded marketing materials, social media posts, and business templates.

3. **Trello** - Project management software helpful for organizing tasks related to production, sales, and marketing.

4. **Mailchimp** - A platform for managing email marketing campaigns essential for customer engagement and retention.

5. **Wave** - A free accounting tool with features like invoicing, expense tracking, and receipt management tailored for small businesses.

Academic Sources on Sustainable and Ethical Practices

1. **"Consumer Trends in Sustainable Food Choices"** - Research published in *Agriculture and Human Values*, 2021, covering the rise of sustainable consumer habits.

2. **USDA Sustainable Agriculture Research and Education (SARE)** - www.sare.org

3.

4. A federal resource offering research articles, guides, and case studies on sustainable agriculture practices.

Also by
M.K. Hanna

Growing Microgreens: From Seed To Table

Made in United States
Troutdale, OR
05/11/2025